Proteomics: Principles, Techniques and Analysis

Proteomics: Principles, Techniques and Analysis

Edited by Peter Wyatt

New York

Published by Syrawood Publishing House,
750 Third Avenue, 9th Floor,
New York, NY 10017, USA
www.syrawoodpublishinghouse.com

Proteomics: Principles, Techniques and Analysis
Edited by Peter Wyatt

© 2018 Syrawood Publishing House

International Standard Book Number: 978-1-68286-597-2 (Hardback)

This book contains information obtained from authentic and highly regarded sources. All chapters are published with permission under the Creative Commons Attribution Share Alike License or equivalent. A wide variety of references are listed. Permissions and sources are indicated; for detailed attributions, please refer to the permissions page. Reasonable efforts have been made to publish reliable data and information, but the authors, editors and publisher cannot assume any responsibility for the vailidity of all materials or the consequences of their use.

Trademark Notice: Registered trademark of products or corporate names are used only for explanation and identification without intent to infringe.

Cataloging-in-Publication Data

Proteomics : principles, techniques and analysis / edited by Peter Wyatt.
   p. cm.
Includes bibliographical references and index.
ISBN 978-1-68286-597-2
1. Proteomics. 2. Proteins. 3. Molecular biology. I. Wyatt, Peter.
QP551 .P76 2018
572.6--dc23

TABLE OF CONTENTS

Preface VII

Chapter 1   **Basics of Proteomics** 1
- Proteomics 1
- Protein 6

Chapter 2   **Protein Breaking and its Application** 29
- Amino Acid 29
- Peptide Bond 47
- Protein Structure 48
- Protein Separation 69

Chapter 3   **Electrophoresis: An Integrated Study** 87
- Electrophoresis 87
- Gel-based Proteomics 90
- Difference in Gel Electrophoresis 106
- Ion Chromatography 110

Chapter 4   **An Overview of Mass Spectrometry** 132
- Mass Spectrometry 132
- Principle of Mass Spectrometer 153
- Hybrid Mass Spectrometer 159

Chapter 5   **Chromatography: A Comprehensive Study** 190
- Chromatography 190
- Affinity Chromatography 202

Permissions

Index

PREFACE

The proteomes are a set of proteins that are produced by an organism. Proteomics helps in understanding the movement and interactions of proteins. Modern technologies have been introduced to develop a better understanding of proteomics. The common techniques used are mass spectrometry, differential in-gel electrophoresis, etc. This textbook, with its detailed analyses and data, will prove immensely beneficial to professionals and students involved in this area at various levels. The topics covered in this book offer the readers new insights in the field of proteomics.

A detailed account of the significant topics covered in this book is provided below:

Chapter 1- Proteomics is devoted to the study of proteins, and the term is a portmanteau of protein and genome. The field normally focuses on experimental analysis of proteins. The term proteome refers to the all types of protein expressed by cells, genomes, tissues etc. This is an introductory chapter which will introduce briefly all the significant aspects of proteomics.

Chapter 2- Techniques are an important component of any field of study. The following section elucidates the various techniques that are related to protein-breaking. To study proteins, it is necessary to separate it from a mixture. Chromatography is a widely used method for separation of proteins. The method can be classified into affinity chromatography, adsorption chromatography, ion-exchange chromatography, partition chromatography, and molecular exclusion (gel filtration) chromatography.

Chapter 3- Science and technology have undergone rapid developments in the past decade which has resulted in the discovery of significant techniques in the separation of proteins. These have been extensively detailed in this chapter. Electrophoresis is another method for separation of protein, which is more powerful and accurate than chromatography. It works on the principle that charged particles move towards oppositely charged electrode in a fluid. This chapter discusses in detail the theories and methodologies related to electrophoresis.

Chapter 4- Mass spectrometry evaluates the mass-to-charge ratio of a particle existing in vacuum. Upon finding the mass-to-charge ratio, particles can be separated, which helps in conducting many experiments. The aspects elucidated in this chapter are of vital importance, and provide a better understanding of mass spectrometry.

Chapter 5- Affinity chromatography falls under liquid chromatography and uses the method of reversible biological interaction. In theoretical terms, this technique can provide complete purification in a single step. Affinity chromatography is best understood in confluence with the major topics listed in the following chapter.

I would like to make a special mention of my publisher who considered me worthy of this opportunity and also supported me throughout the process. I would also like to thank the editing team at the back-end who extended their help whenever required.

Editor

Basics of Proteomics

Proteomics is devoted to the study of proteins, and the term is a portmanteau of protein and genome. The field normally focuses on experimental analysis of proteins. The term proteome refers to the all types of protein expressed by cells, genomes, tissues etc. This is an introductory chapter which will introduce briefly all the significant aspects of proteomics.

Proteomics

The word "proteome" represents the complete protein pool of an organism encoded by the genome. In broader term, Proteomics, is defined as the total protein content of a cell or that of an organism. Proteomics helps in understanding of alteration in protein expression during different stages of life cycle or under stress condition of an organism. Likewise, Proteomics helps in understanding the structure and function of different proteins as well as protein-protein interactions. A minor defect in either protein structure, its function or alternation in expression pattern can be easily detected using proteomics studies. This is important with regards to drug development and understanding various biological processes, as proteins are the most favorable targets for various drugs. Proteomics on the whole can be divided into three kinds as described below:

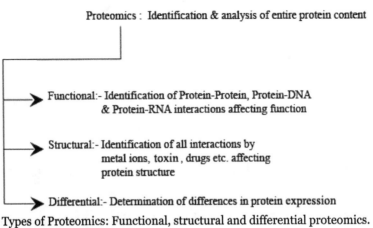

Types of Proteomics: Functional, structural and differential proteomics.

Techniques Involved in Proteomics Study

Some of the very basic analytical techniques are used as major proteomic tools for studying the proteome of an organism. We shall study most of these techniques as we progress in the course. The initial step in all proteomic studies is the separation of a mixture of proteins. This can be carried out using Two Dimensional Gel Electrophoresis technique in which proteins are first of all separated based on their individual charges in 1D. The gel is then turned 90 degrees from its initial position to separate proteins based on the difference in their size. This separation occurs in 2 nd dimension hence the name 2D. The spots obtained in 2D electrophoresis are excised and further subjected to mass spectrometric analysis of each protein present in the mixture.

Apart from charge and size of proteins there are number of other intrinsic properties of proteins that can be employed for their separation and detection. One of these techniques is Field Flow Fractionation (FFF) which separates proteins based on their mobility in presence of applied field. The difference in mobility may be attributed to different size and mass of proteins. The applied field can be of many types such as electrical, gravitational, centrifugal etc. This technique helps in determining different components in a protein mixture, different conformations of protein, their interaction with other proteins as well as some organic molecules such as drugs.

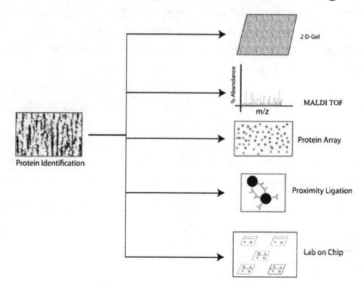

Techniques involved in protein identification during proteomic analysis

Steps in Proteomic Analysis

The following steps are involved in analysis of proteome of an organism as shown in Figure:

1. Purification of proteins: This step involves extraction of protein samples from whole cell, tissue or sub cellular organelles followed by purification using density gradient centrifugation, chromatographic techniques (exclusion, affinity etc.)

2. Separation of proteins: 2D gel electrophoresis is applied for separation of proteins

on the basis of their isoelectric points in one dimension and molecular weight on the other. Spots are detected using fluorescent dyes or radioactive probes.

3. Identification of proteins: The separated protein spots on gel are excised and digested in gel by a protease (e.g. trypsin). The eluted peptides are identified using mass spectrometry.

Analysis of protein molecules is usually carried out by MALDI-TOF (Matrix Assisted Laser Desorption Ionization-Time of Flight) based peptide mass fingerprinting. Determined amino acid sequence is finally compared with available database to validate the proteins.

Several online tools are available for proteomic analysis such as Mascot, Aldente, Popitam, Quickmod, Peptide cutter etc.

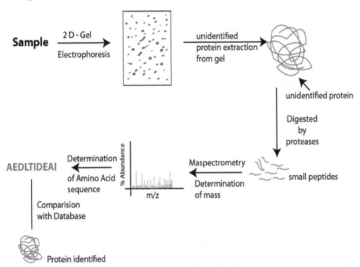

Overview of steps involved in proteomic analysis

Applications of Proteomics

Proteomics has broad applications in all the aspects of life sciences including several practical applications as drug development against several diseases. Difference in expression protein expression profile of normal and diseased person may be analyzed for target protein. Protein to gene may be predicted. Once protein/gene is identified, function may be predicted. This can help in disease management/drug development.

From Protein to Gene

Whole genome sequences of several organisms have been completed but genomic data does not show how proteins function or how these proteins are involved in biological processes. Gene codes for a protein by at several occasion proteins are modified after synthesis (several types of post-translational modifications) for functional diversification.

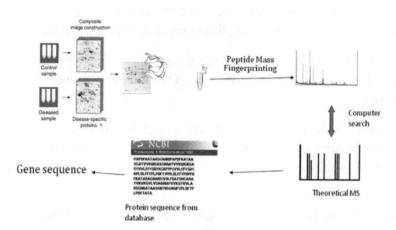

Proteomic analysis-protein to gene sequence

Proteome

The proteome is the entire set of proteins expressed by a genome, cell, tissue, or organism at a certain time. More specifically, it is the set of expressed proteins in a given type of cell or organism, at a given time, under defined conditions. The term is a blend of *proteins* and *genome*. Proteomics is the study of the proteome.

Systems

The term has been applied to several different types of biological systems. A cellular proteome is the collection of proteins found in a particular cell type under a particular set of environmental conditions such as exposure to hormone stimulation. It can also be useful to consider an organism's complete proteome, which can be conceptualized as the complete set of proteins from all of the various cellular proteomes. This is very roughly the protein equivalent of the genome. The term "proteome" has also been used to refer to the collection of proteins in certain sub-cellular biological systems. For example, all of the proteins in a virus can be called a viral proteome.

History

Marc Wilkins coined the term *proteome* in 1994 in a symposium on "2D Electrophoresis: from protein maps to genomes" held in Siena in Italy. It appeared in print in 1995, with the publication of part of Wilkins's PhD thesis. Wilkins used the term to describe the entire complement of proteins expressed by a genome, cell, tissue or organism.

Size and Contents

The proteome can be larger than the genome, especially in eukaryotes, as more than one protein can be produced from one gene due to alternative splicing (e.g. human proteome consists 92,179 proteins out of which 71,173 are splicing variants). On the other hand, not all genes are translated to proteins, and many known genes encode only RNA which is the final functional product. Moreover, complete proteome size vary depending on the kingdom of life. For instance, eukaryotes, bacteria, Archaea and viruses have on average 15145, 3200, 2358 and 42 proteins respectively encoded in their genomes.

Dark Proteome

Perdigão and co-workers surveyed the "dark" proteome – that is, regions of proteins never observed by experimental structure determination and inaccessible to homology modeling. For 546,000 Swiss-Prot proteins, they found that 44–54% of the proteome in eukaryotes and viruses was "dark", compared with only ~14% in archaea and bacteria. Surprisingly, most of the dark proteome could not be accounted for by conventional explanations, such as intrinsic disorder or transmembrane regions. Nearly half of the dark proteome comprised dark proteins, in which the entire sequence lacked similarity to any known structure. Dark proteins fulfill a wide variety of functions, but a subset showed distinct and largely unexpected features, such as association with secretion, specific tissues, the endoplasmic reticulum, disulfide bonding, and proteolytic cleavage.

Methods to Study the Proteome

Numerous methods are available to study proteins, sets of proteins, or the whole proteome. In fact, proteins are often studied indirectly, e.g. using computational methods and analyses of genomes. Only a few examples are given below.

Separation Techniques and Electrophoresis

Proteomics, the study of the proteome, has largely been practiced through the separation of proteins by two dimensional gel electrophoresis. In the first dimension, the proteins are separated by isoelectric focusing, which resolves proteins on the basis of charge. In the second dimension, proteins are separated by molecular weight using SDS-PAGE. The gel is dyed with Coomassie Brilliant Blue or silver to visualize the proteins. Spots on the gel are proteins that have migrated to specific locations.

Mass Spectrometry

Mass spectrometry has augmented proteomics. Peptide mass fingerprinting identifies a protein by cleaving it into short peptides and then deduces the protein's identity by matching the observed peptide masses against a sequence database. Tandem mass spectrometry, on the other hand, can get sequence information from individual peptides by isolating them, colliding them with a non-reactive gas, and then cataloguing the fragment ions produced.

In May 2014, a draft map of the human proteome was published in *Nature*. This map was generated using high-resolution Fourier-transform mass spectrometry. This study profiled 30 histologically normal human samples resulting in the identification of proteins coded by 17,294 genes. This accounts for around 84% of the total annotated protein-coding genes.

Protein Complementation Assays and Interaction Screens

Protein fragment complementation assays are often used to detect protein–protein interactions. The yeast two-hybrid assay is the most popular of them but there are numerous variations, both used *in vitro* and *in vivo*.

Protein

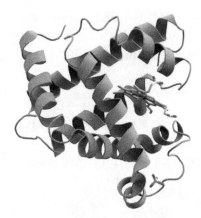

A representation of the 3D structure of the protein myoglobin showing turquoise α-helices. This protein was the first to have its structure solved by X-ray crystallography. Towards the right-center among the coils, a prosthetic group called a heme group (shown in gray) with a bound oxygen molecule (red)

Proteins are large biomolecules, or macromolecules, consisting of one or more long chains of amino acid residues. Proteins perform a vast array of functions within organisms, including catalysing metabolic reactions, DNA replication, responding to stimuli, and transporting molecules from one location to another. Proteins differ from one another primarily in their sequence of amino acids, which is dictated by the nucleotide sequence of their genes, and which usually results in protein folding into a specific three-dimensional structure that determines its activity.

A linear chain of amino acid residues is called a polypeptide. A protein contains at least one long polypeptide. Short polypeptides, containing less than 20–30 residues, are rarely considered to be proteins and are commonly called peptides, or sometimes oligopeptides. The individual amino acid residues are bonded together by peptide bonds and adjacent amino acid residues. The sequence of amino acid residues in a protein is defined by the sequence of a gene, which is encoded in the genetic code. In general, the genetic code specifies 20 standard amino acids; however, in certain organisms the genetic code can include selenocysteine and—in certain archaea—pyrrolysine. Shortly after or even during synthesis, the residues in a protein are often chemically modified by post-translational modification, which alters the physical and chemical properties, folding, stability, activity, and ultimately, the function of the proteins. Sometimes proteins have non-peptide groups attached, which can be called prosthetic groups or cofactors. Proteins can also work together to achieve a particular function, and they often associate to form stable protein complexes.

Once formed, proteins only exist for a certain period of time and are then degraded and recycled by the cell's machinery through the process of protein turnover. A protein's lifespan is measured in terms of its half-life and covers a wide range. They can exist for minutes or years with an average lifespan of 1–2 days in mammalian cells. Abnormal and or misfolded proteins are degraded more rapidly either due to being targeted for destruction or due to being unstable.

Like other biological macromolecules such as polysaccharides and nucleic acids, proteins are essential parts of organisms and participate in virtually every process within cells. Many proteins are enzymes that catalyse biochemical reactions and are vital to metabolism. Proteins also have structural or mechanical functions, such as actin and myosin in muscle and the proteins in the cytoskeleton, which form a system of scaffolding that maintains cell shape. Other proteins are important in cell signaling, immune responses, cell adhesion, and the cell cycle. In animals, proteins are needed in the diet to provide the essential amino acids that cannot be synthesized. Digestion breaks the proteins down for use in the metabolism.

Proteins may be purified from other cellular components using a variety of techniques such as ultracentrifugation, precipitation, electrophoresis, and chromatography; the advent of genetic engineering has made possible a number of methods to facilitate purification. Methods commonly used to study protein structure and function include immunohistochemistry, site-directed mutagenesis, X-ray crystallography, nuclear magnetic resonance and mass spectrometry.

Biochemistry

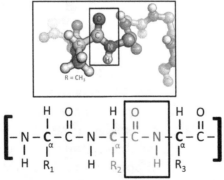

Chemical structure of the peptide bond (bottom) and the three-dimensional structure of a peptide bond between an alanine and an adjacent amino acid (top/inset)

Resonance structures of the peptide bond that links individual amino acids to form a protein polymer

Most proteins consist of linear polymers built from series of up to 20 different L-α-amino acids. All proteinogenic amino acids possess common structural features, including an α-carbon to which an amino group, a carboxyl group, and a variable side chain are bonded. Only proline differs from this basic structure as it contains an unusual ring to the N-end amine group, which forces the CO–NH amide moiety into a fixed conformation. The side chains of the standard amino acids, detailed in the list of standard amino acids, have a great variety of chemical structures and properties; it is the combined effect of all of the amino acid side chains in a protein that ultimately determines its three-dimensional structure and its chemical reactivity. The amino acids in a polypeptide chain are linked by peptide bonds. Once linked in the protein chain, an individual amino acid is called a *residue,* and the linked series of carbon, nitrogen, and oxygen atoms are known as the *main chain* or *protein backbone.*

The peptide bond has two resonance forms that contribute some double-bond character and inhibit rotation around its axis, so that the alpha carbons are roughly coplanar. The other two dihedral angles in the peptide bond determine the local shape assumed by the protein backbone. The end with a free amino group is known as the N-terminus or amino terminus, whereas the end of the protein with a free carboxyl group is known as the C-terminus or carboxy terminus (the sequence of the protein is written from N-terminus to C-terminus, from left to right).

The words *protein, polypeptide,* and *peptide* are a little ambiguous and can overlap in meaning. *Protein* is generally used to refer to the complete biological molecule in a stable conformation, whereas *peptide* is generally reserved for a short amino acid oligomers often lacking a stable three-dimensional structure. However, the boundary between the two is not well defined and usually lies near 20–30 residues. *Polypeptide* can refer to any single linear chain of amino acids, usually regardless of length, but often implies an absence of a defined conformation.

Abundance in Cells

It has been estimated that average-sized bacteria contain about 2 million proteins per cell (e.g. *E. coli* and *Staphylococcus aureus*). Smaller bacteria, such as *Mycoplasma* or *spirochetes* contain fewer molecules, namely on the order of 50,000 to 1 million. By contrast, eukaryotic cells are larger and thus contain much more protein. For instance, yeast cells were estimated to contain about 50 million proteins and human cells on the order of 1 to 3 billion. The concentration of individual protein copies ranges from a few molecules per cell up to 20 million. Not all genes coding proteins are expressed in most cells and their number depends on for example cell type and external stimuli. For instance, of the 20,000 or so proteins encoded by the human genome, only 6,000 are detected in lymphoblastoid cells. Moreover, the number of proteins the genome encodes correlates well with the organism complexity. Eukaryotes, bacteria, Archaea and viruses have on average 15145, 3200, 2358 and 42 proteins respectively coded in their genomes.

Synthesis

Biosynthesis

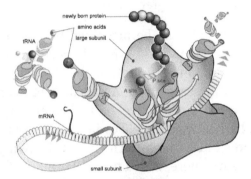

A ribosome produces a protein using mRNA as template

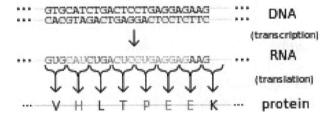

The DNA sequence of a gene encodes the amino acid sequence of a protein

Proteins are assembled from amino acids using information encoded in genes. Each protein has its own unique amino acid sequence that is specified by the nucleotide sequence of the gene encoding this protein. The genetic code is a set of three-nucleotide sets called codons and each three-nucleotide combination designates an amino acid, for example AUG (adenine-uracil-guanine) is the code for methionine. Because DNA contains four nucleotides, the total number of possible codons is 64; hence, there is some redundancy in the genetic code, with some amino acids specified by more than one codon. Genes encoded in DNA are first transcribed into pre-messenger RNA (mRNA) by proteins such as RNA polymerase. Most organisms then process the pre-mRNA (also known as a *primary transcript*) using various forms of Post-transcriptional modification to form the mature mRNA, which is then used as a template for protein synthesis by the ribosome. In prokaryotes the mRNA may either be used as soon as it is produced, or be bound by a ribosome after having moved away from the nucleoid. In contrast, eukaryotes make mRNA in the cell nucleus and then translocate it across the nuclear membrane into the cytoplasm, where protein synthesis then takes place. The rate of protein synthesis is higher in prokaryotes than eukaryotes and can reach up to 20 amino acids per second.

The process of synthesizing a protein from an mRNA template is known as translation. The mRNA is loaded onto the ribosome and is read three nucleotides at a time by matching each codon to its base pairing anticodon located on a transfer RNA molecule, which carries the amino acid corresponding to the codon it recognizes. The enzyme aminoacyl tRNA synthetase "charges" the tRNA molecules with the correct amino acids. The growing polypeptide is often termed the *nascent chain*. Proteins are always biosynthesized from N-terminus to C-terminus.

The size of a synthesized protein can be measured by the number of amino acids it contains and by its total molecular mass, which is normally reported in units of *daltons* (synonymous with atomic mass units), or the derivative unit kilodalton (kDa). The average size of protein increases from Archaea, Bacteria to Eukaryote (283, 311, 438 residues and 31, 34, 49 kDa respecitvely) due bigger number of protein domains constituting proteins in higher organisms. For instance, yeast proteins are on average 466 amino acids long and 53 kDa in mass. The largest known proteins are the titins, a component of the muscle sarcomere, with a molecular mass of almost 3,000 kDa and a total length of almost 27,000 amino acids.

Chemical Synthesis

Short proteins can also be synthesized chemically by a family of methods known as peptide synthesis, which rely on organic synthesis techniques such as chemical ligation to produce peptides in high yield. Chemical synthesis allows for the introduction of non-natural amino acids into polypeptide chains, such as attachment of fluorescent probes to amino acid side chains. These methods are useful in laboratory biochemistry and cell biology, though generally not for commercial applications. Chemical synthesis is inefficient for polypeptides longer than about 300 amino acids, and the synthesized proteins may not readily assume their native tertiary structure. Most chemical synthesis methods proceed from C-terminus to N-terminus, opposite the biological reaction.

Structure

The crystal structure of the chaperonin, a huge protein complex. A single protein subunit is highlighted. Chaperonins assist protein folding

Three possible representations of the three-dimensional structure of the protein triose phosphate isomerase. Left: All-atom representation colored by atom type. Middle: Simplified representation illustrating the backbone conformation, colored by secondary structure. Right: Solvent-accessible surface representation colored by residue type (acidic residues red, basic residues blue, polar residues green, nonpolar residues white)

Most proteins fold into unique 3-dimensional structures. The shape into which a protein naturally folds is known as its native conformation. Although many proteins can fold unassisted, simply through the chemical properties of their amino acids, others require the aid of molecular chaperones to fold into their native states. Biochemists often refer to four distinct aspects of a protein's structure:

- *Primary structure*: the amino acid sequence. A protein is a polyamide.

- *Secondary structure*: regularly repeating local structures stabilized by hydrogen bonds. The most common examples are the α-helix, β-sheet and turns. Because secondary structures are local, many regions of different secondary structure can be present in the same protein molecule.

- *Tertiary structure*: the overall shape of a single protein molecule; the spatial relationship of the secondary structures to one another. Tertiary structure is generally stabilized by nonlocal interactions, most commonly the formation of a hydrophobic core, but also through salt bridges, hydrogen bonds, disulfide bonds, and even posttranslational modifications. The term "tertiary structure" is often used as synonymous with the term *fold*. The tertiary structure is what controls the basic function of the protein.

- *Quaternary structure*: the structure formed by several protein molecules (polypeptide chains), usually called *protein subunits* in this context, which function as a single protein complex.

Proteins are not entirely rigid molecules. In addition to these levels of structure, proteins may shift between several related structures while they perform their functions. In the context of these functional rearrangements, these tertiary or quaternary structures are usually referred to as "conformations", and transitions between them are called *conformational changes*. Such changes are often induced by the binding of a substrate molecule to an enzyme's active site, or the physical region of the protein that participates in chemical catalysis. In solution proteins also undergo variation in structure through thermal vibration and the collision with other molecules.

Molecular surface of several proteins showing their comparative sizes. From left to right are: immunoglobulin G (IgG, an antibody), hemoglobin, insulin (a hormone), adenylate kinase (an enzyme), and glutamine synthetase (an enzyme)

Proteins can be informally divided into three main classes, which correlate with typical tertiary structures: globular proteins, fibrous proteins, and membrane proteins. Almost all globular proteins are soluble and many are enzymes. Fibrous proteins are often structural, such as collagen, the major component of connective tissue, or keratin, the protein component of hair and nails. Membrane proteins often serve as receptors or provide channels for polar or charged molecules to pass through the cell membrane.

A special case of intramolecular hydrogen bonds within proteins, poorly shielded from water attack and hence promoting their own dehydration, are called dehydrons.

Structure Determination

Discovering the tertiary structure of a protein, or the quaternary structure of its complexes, can provide important clues about how the protein performs its function. Common experimental methods of structure determination include X-ray crystallography and NMR spectroscopy, both of which can produce information at atomic resolution. However, NMR experiments are able to provide information from which a subset of distances between pairs of atoms can be estimated, and the final possible conformations for a protein are determined by solving a distance geometry problem. Dual polarisation interferometry is a quantitative analytical method for measuring the overall protein conformation and conformational changes due to interactions or other stimulus. Circular dichroism is another laboratory technique for determining internal β-sheet / α-helical composition of proteins. Cryoelectron microscopy is used to produce lower-resolution structural information about very large protein complexes, including assembled viruses; a variant known as electron crystallography can also produce high-resolution information in some cases, especially for two-dimensional crystals of membrane proteins. Solved structures are usually deposited in the Protein Data Bank (PDB), a freely available resource from which structural data about thousands of proteins can be obtained in the form of Cartesian coordinates for each atom in the protein.

Many more gene sequences are known than protein structures. Further, the set of solved structures is biased toward proteins that can be easily subjected to the conditions required in X-ray crystallography, one of the major structure determination methods. In particular, globular proteins are comparatively easy to crystallize in preparation for X-ray crystallography. Membrane proteins, by contrast, are difficult to crystallize and are underrepresented in the PDB. Structural genomics initiatives have attempted to remedy these deficiencies by systematically solving representative structures of major fold classes. Protein structure prediction methods attempt to provide a means of generating a plausible structure for proteins whose structures have not been experimentally determined.

Cellular Functions

Proteins are the chief actors within the cell, said to be carrying out the duties specified by the information encoded in genes. With the exception of certain types of RNA, most other biological molecules are relatively inert elements upon which proteins act. Proteins make up half the dry weight of an *Escherichia coli* cell, whereas other macromolecules such as DNA and RNA make up only 3% and 20%, respectively. The set of proteins expressed in a particular cell or cell type is known as its proteome.

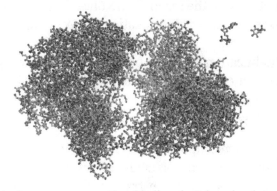

The enzyme hexokinase is shown as a conventional ball-and-stick molecular model. To scale in the top right-hand corner are two of its substrates, ATP and glucose

The chief characteristic of proteins that also allows their diverse set of functions is their ability to bind other molecules specifically and tightly. The region of the protein responsible for binding another molecule is known as the binding site and is often a depression or "pocket" on the molecular surface. This binding ability is mediated by the tertiary structure of the protein, which defines the binding site pocket, and by the chemical properties of the surrounding amino acids' side chains. Protein binding can be extraordinarily tight and specific; for example, the ribonuclease inhibitor protein binds to human angiogenin with a sub-femtomolar dissociation constant ($<10^{-15}$ M) but does not bind at all to its amphibian homolog onconase (>1 M). Extremely minor chemical changes such as the addition of a single methyl group to a binding partner can sometimes suffice to nearly eliminate binding; for example, the aminoacyl tRNA synthetase specific to the amino acid valine discriminates against the very similar side chain of the amino acid isoleucine.

Proteins can bind to other proteins as well as to small-molecule substrates. When proteins bind specifically to other copies of the same molecule, they can oligomerize to form fibrils; this process occurs often in structural proteins that consist of globular monomers that self-associate to form rigid fibers. Protein–protein interactions also regulate enzymatic activity, control progression through the cell cycle, and allow the assembly of large protein complexes that carry out many closely related reactions with a common biological function. Proteins can also bind to, or even be integrated into, cell membranes. The ability of binding partners to induce conformational changes in proteins allows the construction of enormously complex signaling networks. As interactions between proteins are reversible, and depend heavily on the availability of different groups of partner proteins to form aggregates that are capable to carry out discrete sets of function, study of the interactions between specific proteins is a key to understand important aspects of cellular function, and ultimately the properties that distinguish particular cell types.

Enzymes

The best-known role of proteins in the cell is as enzymes, which catalyse chemical reactions. Enzymes are usually highly specific and accelerate only one or a few chemical reactions. Enzymes carry out most of the reactions involved in metabolism, as well as manipulating DNA in processes such as DNA replication, DNA repair, and transcription. Some enzymes act on other proteins to add or remove chemical groups in a process known as posttranslational modification. About 4,000 reactions are known to be catalysed by enzymes. The rate acceleration conferred by enzymatic catalysis is often enormous—as much as 10^{17}-fold increase in rate over the uncatalysed reaction in the case of orotate decarboxylase (78 million years without the enzyme, 18 milliseconds with the enzyme).

The molecules bound and acted upon by enzymes are called substrates. Although enzymes can consist of hundreds of amino acids, it is usually only a small fraction of the residues that come in contact with the substrate, and an even smaller fraction—three to four residues on average—that are directly involved in catalysis. The region of the enzyme that binds the substrate and contains the catalytic residues is known as the active site.

Dirigent proteins are members of a class of proteins that dictate the stereochemistry of a compound synthesized by other enzymes.

Cell Signaling And Ligand Binding

Ribbon diagram of a mouse antibody against cholera that binds a carbohydrate antigen

Many proteins are involved in the process of cell signaling and signal transduction. Some proteins, such as insulin, are extracellular proteins that transmit a signal from the cell in which they were synthesized to other cells in distant tissues. Others are membrane proteins that act as receptors whose main function is to bind a signaling molecule and induce a biochemical response in the cell. Many receptors have a binding site exposed on the cell surface and an effector domain within the cell, which may have enzymatic activity or may undergo a conformational change detected by other proteins within the cell.

Antibodies are protein components of an adaptive immune system whose main function is to bind antigens, or foreign substances in the body, and target them for destruction. Antibodies can be secreted into the extracellular environment or anchored in the membranes of specialized B cells known as plasma cells. Whereas enzymes are limited in their binding affinity for their substrates by the necessity of conducting their reaction, antibodies have no such constraints. An antibody's binding affinity to its target is extraordinarily high.

Many ligand transport proteins bind particular small biomolecules and transport them to other locations in the body of a multicellular organism. These proteins must have a high binding affinity when their ligand is present in high concentrations, but must also release the ligand when it is present at low concentrations in the target tissues. The canonical example of a ligand-binding protein is haemoglobin, which transports oxygen from the lungs to other organs and tissues in all vertebrates and has close homologs in every biological kingdom. Lectins are sugar-binding proteins which are highly specific for their sugar moieties. Lectins typically play a role in biological recognition phenomena involving cells and proteins. Receptors and hormones are highly specific binding proteins.

Transmembrane proteins can also serve as ligand transport proteins that alter the permeability of the cell membrane to small molecules and ions. The membrane alone has a hydrophobic core through which polar or charged molecules cannot diffuse. Membrane proteins contain internal channels that allow such molecules to enter and exit the cell. Many ion channel proteins are specialized to select for only a particular ion; for example, potassium and sodium channels often discriminate for only one of the two ions.

Structural Proteins

Structural proteins confer stiffness and rigidity to otherwise-fluid biological components. Most structural proteins are fibrous proteins; for example, collagen and elastin are critical components of connective tissue such as cartilage, and keratin is found in hard or filamentous structures such as hair, nails, feathers, hooves, and some animal shells. Some globular proteins can also play structural functions, for example, actin and tubulin are globular and soluble as monomers, but polymerize to form long, stiff fibers that make up the cytoskeleton, which allows the cell to maintain its shape and size.

Other proteins that serve structural functions are motor proteins such as myosin, kinesin, and dynein, which are capable of generating mechanical forces. These proteins are crucial for cellular motility of single celled organisms and the sperm of many multicellular organisms which reproduce sexually. They also generate the forces exerted by contracting muscles and play essential roles in intracellular transport.

Methods of Study

The activities and structures of proteins may be examined *in vitro, in vivo, and in silico*. *In vitro* studies of purified proteins in controlled environments are useful for learning how a protein carries out its function: for example, enzyme kinetics studies explore the chemical mechanism of an enzyme's catalytic activity and its relative affinity for various possible substrate molecules. By contrast, *in vivo* experiments can provide information about the physiological role of a protein in the context of a cell or even a whole organism. *In silico* studies use computational methods to study proteins.

Protein Purification

To perform *in vitro* analysis, a protein must be purified away from other cellular components. This process usually begins with cell lysis, in which a cell's membrane is disrupted and its internal contents released into a solution known as a crude lysate. The resulting mixture can be purified using ultracentrifugation, which fractionates the various cellular components into fractions containing soluble proteins; membrane lipids and proteins; cellular organelles, and nucleic acids. Precipitation by a method known as salting out can concentrate the proteins from this lysate. Various types of chromatography are then used to isolate the protein or proteins of interest based on properties such as molecular weight, net charge and binding affinity. The level of purification can be monitored using various types of gel electrophoresis if the desired protein's molecular weight and isoelectric point are known, by spectroscopy if the protein has distinguishable spectroscopic features, or by enzyme assays if the protein has enzymatic activity. Additionally, proteins can be isolated according their charge using electrofocusing.

For natural proteins, a series of purification steps may be necessary to obtain protein sufficiently pure for laboratory applications. To simplify this process, genetic engineering is often used to add chemical features to proteins that make them easier to purify without affecting their structure or activity. Here, a "tag" consisting of a specific amino acid sequence, often a series of histidine residues (a "His-tag"), is attached to one terminus of the protein. As a result, when the lysate is passed over a chromatography column containing nickel, the histidine residues ligate the nickel and attach to the column while the untagged components of the lysate pass unimpeded. A number of different tags have been developed to help researchers purify specific proteins from complex mixtures.

Basics of Proteomics

Cellular Localization

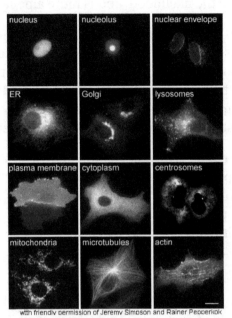

with friendly permission of Jeremy Simpson and Rainer Pepperkok

Proteins in different cellular compartments and structures tagged with green fluorescent protein (here, white)

The study of proteins *in vivo* is often concerned with the synthesis and localization of the protein within the cell. Although many intracellular proteins are synthesized in the cytoplasm and membrane-bound or secreted proteins in the endoplasmic reticulum, the specifics of how proteins are targeted to specific organelles or cellular structures is often unclear. A useful technique for assessing cellular localization uses genetic engineering to express in a cell a fusion protein or chimera consisting of the natural protein of interest linked to a "reporter" such as green fluorescent protein (GFP). The fused protein's position within the cell can be cleanly and efficiently visualized using microscopy, as shown in the figure opposite.

Other methods for elucidating the cellular location of proteins requires the use of known compartmental markers for regions such as the ER, the Golgi, lysosomes or vacuoles, mitochondria, chloroplasts, plasma membrane, etc. With the use of fluorescently tagged versions of these markers or of antibodies to known markers, it becomes much simpler to identify the localization of a protein of interest. For example, indirect immunofluorescence will allow for fluorescence colocalization and demonstration of location. Fluorescent dyes are used to label cellular compartments for a similar purpose.

Other possibilities exist, as well. For example, immunohistochemistry usually utilizes an antibody to one or more proteins of interest that are conjugated to enzymes yielding either luminescent or chromogenic signals that can be compared between samples, allowing for localization information. Another applicable technique is cofractionation in sucrose (or other material) gradients using isopycnic centrifugation. While this technique does not prove colocalization of a compartment of known density and the protein of interest, it does increase the likelihood, and is more amenable to large-scale studies.

Finally, the gold-standard method of cellular localization is immunoelectron microscopy. This technique also uses an antibody to the protein of interest, along with classical electron microscopy techniques. The sample is prepared for normal electron microscopic examination, and then treated with an antibody to the protein of interest that is conjugated to an extremely electro-dense material, usually gold. This allows for the localization of both ultrastructural details as well as the protein of interest.

Through another genetic engineering application known as site-directed mutagenesis, researchers can alter the protein sequence and hence its structure, cellular localization, and susceptibility to regulation. This technique even allows the incorporation of unnatural amino acids into proteins, using modified tRNAs, and may allow the rational design of new proteins with novel properties.

Proteomics

The total complement of proteins present at a time in a cell or cell type is known as its proteome, and the study of such large-scale data sets defines the field of proteomics, named by analogy to the related field of genomics. Key experimental techniques in proteomics include 2D electrophoresis, which allows the separation of a large number of proteins, mass spectrometry, which allows rapid high-throughput identification of proteins and sequencing of peptides (most often after in-gel digestion), protein microarrays, which allow the detection of the relative levels of a large number of proteins present in a cell, and two-hybrid screening, which allows the systematic exploration of protein–protein interactions. The total complement of biologically possible such interactions is known as the interactome. A systematic attempt to determine the structures of proteins representing every possible fold is known as structural genomics.

Bioinformatics

A vast array of computational methods have been developed to analyze the structure, function, and evolution of proteins.

The development of such tools has been driven by the large amount of genomic and proteomic data available for a variety of organisms, including the human genome. It is simply impossible to study all proteins experimentally, hence only a few are subjected to laboratory experiments while computational tools are used to extrapolate to similar proteins. Such homologous proteins can be efficiently identified in distantly related organisms by sequence alignment. Genome and gene sequences can be searched by a variety of tools for certain properties. Sequence profiling tools can find restriction enzyme sites, open reading frames in nucleotide sequences, and predict secondary structures. Phylogenetic trees can be constructed and evolutionary hypotheses developed using special software like ClustalW regarding the ancestry of modern organisms and the genes they express. The field of bioinformatics is now indispensable for the analysis of genes and proteins.

Structure Prediction and Simulation

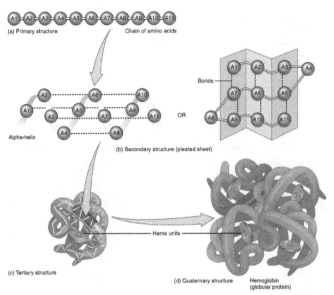

Constituent amino-acids can be analyzed to predict secondary, tertiary and quaternary protein structure, in this case hemoglobin containing heme units

Complementary to the field of structural genomics, *protein structure prediction* develops efficient mathematical models of proteins to computationally predict their structures in theory, instead of detecting structures with laboratory observation. The most successful type of structure prediction, known as homology modeling, relies on the existence of a "template" structure with sequence similarity to the protein being modeled; structural genomics' goal is to provide sufficient representation in solved structures to model most of those that remain. Although producing accurate models remains a challenge when only distantly related template structures are available, it has been suggested that sequence alignment is the bottleneck in this process, as quite accurate models can be produced if a "perfect" sequence alignment is known. Many structure prediction methods have served to inform the emerging field of protein engineering, in which novel protein folds have already been designed. A more complex computational problem is the prediction of intermolecular interactions, such as in molecular docking and protein–protein interaction prediction.

Mathematical models to simulate dynamic processes of protein folding and binding involve molecular mechanics, in particular, molecular dynamics. Monte Carlo techniques facilitate the computations, which exploit advances in parallel and distributed computing (for example, the Folding@home project which performs molecular modeling on GPUs). *In silico* simulations discovered the folding of small α-helical protein domains such as the villin headpiece and the HIV accessory protein. Hybrid methods combining standard molecular dynamics with quantum mechanical mathematics explored the electronic states of rhodopsins.

Protein Disorder and Unstructure Prediction

Many proteins (in Eucaryota ~33%) contain large unstructured but biologically functional segments and can be classified as intrinsically disordered proteins. Predicting and analysing protein disorder is, therefore, an important part of protein structure characterisation.

Nutrition

Most microorganisms and plants can biosynthesize all 20 standard amino acids, while animals (including humans) must obtain some of the amino acids from the diet. The amino acids that an organism cannot synthesize on its own are referred to as essential amino acids. Key enzymes that synthesize certain amino acids are not present in animals — such as aspartokinase, which catalyses the first step in the synthesis of lysine, methionine, and threonine from aspartate. If amino acids are present in the environment, microorganisms can conserve energy by taking up the amino acids from their surroundings and downregulating their biosynthetic pathways.

In animals, amino acids are obtained through the consumption of foods containing protein. Ingested proteins are then broken down into amino acids through digestion, which typically involves denaturation of the protein through exposure to acid and hydrolysis by enzymes called proteases. Some ingested amino acids are used for protein biosynthesis, while others are converted to glucose through gluconeogenesis, or fed into the citric acid cycle. This use of protein as a fuel is particularly important under starvation conditions as it allows the body's own proteins to be used to support life, particularly those found in muscle.

History and Etymology

Proteins were recognized as a distinct class of biological molecules in the eighteenth century by Antoine Fourcroy and others, distinguished by the molecules' ability to coagulate or flocculate under treatments with heat or acid. Noted examples at the time included albumin from egg whites, blood serum albumin, fibrin, and wheat gluten.

Proteins were first described by the Dutch chemist Gerardus Johannes Mulder and named by the Swedish chemist Jöns Jacob Berzelius in 1838. Mulder carried out elemental analysis of common proteins and found that nearly all proteins had the same empirical formula, $C_{400}H_{620}N_{100}O_{120}P_1S_1$. He came to the erroneous conclusion that they might be composed of a single type of (very large) molecule. The term "protein" to describe these molecules was proposed by Mulder's associate Berzelius;

protein is derived from the Greek word (*proteios*), meaning "primary", "in the lead", or "standing in front", + -*in*. Mulder went on to identify the products of protein degradation such as the amino acid leucine for which he found a (nearly correct) molecular weight of 131 Da.

Early nutritional scientists such as the German Carl von Voit believed that protein was the most important nutrient for maintaining the structure of the body, because it was generally believed that "flesh makes flesh." Karl Heinrich Ritthausen extended known protein forms with the identification of glutamic acid. At the Connecticut Agricultural Experiment Station a detailed review of the vegetable proteins was compiled by Thomas Burr Osborne. Working with Lafayette Mendel and applying Liebig's law of the minimum in feeding laboratory rats, the nutritionally essential amino acids were established. The work was continued and communicated by William Cumming Rose. The understanding of proteins as polypeptides came through the work of Franz Hofmeister and Hermann Emil Fischer in 1902. The central role of proteins as enzymes in living organisms was not fully appreciated until 1926, when James B. Sumner showed that the enzyme urease was in fact a protein.

The difficulty in purifying proteins in large quantities made them very difficult for early protein biochemists to study. Hence, early studies focused on proteins that could be purified in large quantities, e.g., those of blood, egg white, various toxins, and digestive/metabolic enzymes obtained from slaughterhouses. In the 1950s, the Armour Hot Dog Co. purified 1 kg of pure bovine pancreatic ribonuclease A and made it freely available to scientists; this gesture helped ribonuclease A become a major target for biochemical study for the following decades.

John Kendrew with model of myoglobin in progress

Linus Pauling is credited with the successful prediction of regular protein secondary structures based on hydrogen bonding, an idea first put forth by William Astbury in 1933. Later work by Walter Kauzmann on denaturation, based partly on previous studies by Kaj Linderstrøm-Lang, contributed an understanding of protein folding and structure mediated by hydrophobic interactions.

The first protein to be sequenced was insulin, by Frederick Sanger, in 1949. Sanger correctly determined the amino acid sequence of insulin, thus conclusively demonstrating that proteins consisted of linear polymers of amino acids rather than branched chains, colloids, or cyclols. He won the Nobel Prize for this achievement in 1958.

The first protein structures to be solved were hemoglobin and myoglobin, by Max Perutz and Sir John Cowdery Kendrew, respectively, in 1958. As of 2017, the Protein Data Bank has over 126,060 atomic-resolution structures of proteins. In more recent times, cryo-electron microscopy of large macromolecular assemblies and computational protein structure prediction of small protein domains are two methods approaching atomic resolution.

Sample Preparation in Proteomics

Sample preparation is by far the most important step in proteomics. The quality of protein sample is directly proportional to the quality of proteomic data that is obtained and hence the interpretation. The overall aim of sample preparation is to obtain a pure protein pool, devoid of any salt or other bio-molecule contamination. In any experiment, reproducibility of data reinforces the hypothesis proposed and hence for this purpose, the sample preparation is considered as one of the most important steps in proteomics. Proteomic samples range from cells (bacterial or yeast) to tissues (plant or animal) to organism as a whole to biofluids (cerebrospinal fluid or serum). With the ultimate aim of extracting proteins from the samples, all protocols need to be optimized accordingly. Sample preparation basically involves solubilizing the proteins followed by their denaturation, reduction and reconstitution in suitable buffer. The global protein extraction process ensures that all cytosolic and organellar proteins are extracted and the interfering compounds are removed. For example, in case of bacterial protein extraction, a gentle cell lysis results into the liberation of all cellular proteins into the medium, whereas in case of plant cells, a rigorous cell lysis method results in the liberation of only cytosolic proteins into the medium. Further, ultracentrifugation is required to separate out the organelles and their proteins. Hence, it is the sample preparation, which determines the fate of the proteins.

Generalized Protein Preparation Protocol

A typical protein isolation protocol involves the basic steps: Solubilize proteins, prevent protein aggregation, denature and reduce all the proteins, remove nucleic acid and other contaminations. A typical protein pool consists of various kinds of proteins varying in their physico-chemical properties. Therefore, a medium should be selected such that it is able to solubilize all the proteins and also prevents their aggregation. The proteomic techniques involved in the experiments are highly sensitive to contamination from salts or nucleic acids or any other bio-molecules. Also, repeated freeze thawing should be avoided and hence, the protein samples before processing are stored at -200C for long time. Usually buffer containing strong chaotropic agents like urea, thiourea and detergents like CHAPS are preferred for solubilizing the proteins. The quality of the sample is much more important than the quantity itself, as quantity can be compromised by pooling in samples together, but quality cannot be compromised as the errors are multiplicative in nature and the error made in the first step of protein isolation gets multiplied and pose a bigger threat to the downstream analysis of data.

Workflow of Sample Preparation

Usually sample preparation starts with disruption of cells or removal of contaminants. Ideally, all the protein extraction steps should be performed at 40C to prevent proteolysis by addition of protease inhibitors. Centrifugation steps and precipitation steps should be carried out for adequate time points to prevent precipitation of unwanted substances.

Cell Disruption / Lysis

Cell disruption or lysis is the first step towards any sample preparation, be it for proteins or DNA or RNA. The level of strain that a cell can handle governs the cell lysis protocol. For example, a fragile eukaryotic animal cell can be effectively lysed enzymatically or by gentle sonication. However, a vigorous sonication is required for bacterial or yeast cells containing cell wall, while a liquid nitrogen crushing is required for algal or plant cell materials. The purpose of cell lysis is to make the proteins and all the bio-molecules available to the external media. Hence, according to the cell type, lysis steps are employed, which may be harsh or gentle or a mixture of both.

The gentle lysis of cells can be performed with the help of agents like osmotic stress (when placed in a hypotonic solution, the cell swells and bursts), detergents (solubilizing the membranes), enzymes (degrading cell walls mainly) and rapid freeze thawing (crystals generated pierces the cells). The cell types which are usually subjected to gentle lysis are mainly tissue culture cells, blood cells, bacterial cells etc. The harsh lysis involves procedures like sonication, French press, manual and mechanical grinders. Usually, plant and algal cells having a tough cell wall are subjected to such treatments or combination of treatments. Sonication and liquid nitrogen crushing appear to be the best protocols for cell lysis in case of plant cells. Over-sonication leads to fragmentation of proteins as well as the DNA which when sheared improperly, leads to increase in the viscocity of the medium. Another problem which arises in liquid nitrogen crushing, is the space generated between the pestle and the cellular paste. To avoid this issue the acid treated sand or glass beads can be added, which further helps in grinding. In majority of the cases, especially when dealing with tough tissues or cells, a combination of more than one method is applied.

Protection From Proteolysis

Once the cells are lysed, all the ingredients come into the medium. Proteases are enzymes, which fragment the proteins into smaller pieces. At normal room temperature, proteases are extremely active and if cell lysis takes place at the room temperature it results into the enormous pool of proteases. To prevent this, either cell lysis is carried out at 4OC or protease inhibitor cocktails are added to the medium prior to cell lysis. Proteases fall under three broader categories – serine proteases, cysteine protease and metalloprotease. A protease inhibitor cocktail usually consists of chemicals that target these key residues of the protease. Apart from the proteases, phosphatases are another set of enzymes, which pose a threat to the proteins, especially, when studying post-translational modifications and signaling is a question. Hence, along with protease inhibitors, phosphatase inhibitors like sodium orthovandate are also added.

Sample Fractionation

A crude protein pool consists of proteins of varying concentrations. In fact the orders of concentration of proteins may vary upto 7- 8 orders of magnitude. Hence it becomes necessary to fractionate them so that a complete representation of the entire proteome pool is available in the sample. For example, in serum, abundant proteins such as albumin, immunoglobulin G and transferrin account for approximately 90% of the total proteins. The low molecular weight proteins either get masked under these abundant proteins or are extremely low in amount to be detected by any technique. Hence, there is a requirement of a suitable pre-fractionation step, which will remove all the unwanted proteins, either completely or partially, and thereby enriching the sample with other proteins. Majority of the pre-fractionation methods are based on selective removal of abundant proteins from complex mixtures and hence affinity chromatography or immunoprecipitation based methods are widely used.

Protein Extraction and Solubilization

Protein extraction is usually performed using organic solvents like acetone or trichloro-acetic acid. The organic solvents are responsible for increasing the protein-protein interaction by removing the solvation spheres around the proteins. As more and more proteins interact, they aggregate and hence precipitate down. For solubilizing the proteins, chaotropic agents Urea, thiourea and detergents such as CHAPS are used. The chaotropic agents and surfactants are responsible for denaturing the proteins, which further break their intermolecular and intramolecular interactions. Once these interactions are broken, the water molecules are able to solubilize the proteins much easily. The proteins are first separated from nucleic acids and membranes by suitable treatments and then precipitated from the solution by use of organic solvents. In certain cases, especially where membrane proteins are involved, their greater hydrophobicity prevent them from easy dissolution. Thio-urea in this aspect becomes extremely helpful, in terms of stabilizing the membrane proteins. In many cases, ionic detergents like SDS pose a problem in downstream processes like electrophoresis. Zwitter-ionic detergents such as CHAPS (3-((3-cholamidopropyl) dimethylamino)-1-propane sulfonate) are extremely useful in solubilization and are also compatible with electrophoresis. Addition of DTT (dithiothreitol) enhances solubilization by reducing the disulfide linkages. Even though acetone and TCA are effective precipitants, care should be taken during removal of excess solvents from the precipitants. The sample should also not be extremely dry, or else, solubiliziation in appropriate buffer becomes an issue.

Contaminant Removal

Contaminants in the form of salts and nucleic acids are the predominant species that need to be removed to obtain high quality protein samples. Usually in TCA Acetone mode of cell lysis and precipitation, the DNA associated with the proteins also precipitate and hence clogs the gel pores during electrophoresis. Hence, the lysis buffer should contain nucleases to digest the DNA and also reduce the sample viscocity. Salts are extremely notorious contaminants in protein solution. They mainly interfere with the isoelectric focusing by allowing the passage of excess current and hence generating substantial heat to destroy the gels and the samples. When the samples are body fluids like serum, plasma or urine, the concentration of salts are very high and hence additional steps for desalting needs to be performed. The small carbon chain length containing columns are commercially available, which by virtue of hydrophobic interactions; adsorb the proteins allowing the salts to pass through them. The proteins are then selectively eluted using formic acid and acetonitrile solution. This approach is usually done before mass spectrometry analysis of proteins because mass spectrometer is extremely sensitive to high concentrations of salt.

Salt removal techniques usually involve dialysis, precipitation and resolubilization, gel filtration etc. But these processes results into diluting proteins and loss of protein quantity; therefore, now a days, commercial zip tipping columns are available for desalting purposes. Polysaccharides and lipids fall under the category of lower level contaminants. Usually they don't pose much problem except when they are in higher amounts, when they tend to clog the gel pores in electrophoresis. Polysaccharides and lipids are highly soluble in organic solvents and hence during the process of protein precipitation, they are easily removed. Pigments like carotenoids or chlorophylls need an additional step of removal by organic solvents such as methanol and chloroform.

Quantification

Quantification is an important step because it governs how much protein is analyzed in a proteomics experiment. Varying amounts of protein loading gives misleading results, especially when the purpose is to study the expression levels of proteins. Various methods of protein quantification are available. The most common and traditional method involves measuring the absorbance at 280 nm, identifying the aromatic amino acids like tyrosine, tryptophan and phenylalanine. However, it is a highly error prone assay with a biasness for aromatic amino acids. Any protein having an under-representation of aromatic amino acid is thus lost during the process of quantification.

Dye based assays like Lowry, Bradford or Bicinchoninic acid are much preferred over the conventional UV based assay. Lowry assay involves the combination of Biuret Assay involving reduction of cupric ions to cuprous ions and the stabilization of the complex by alkaline tartarate. The reaction yields a purple colour and the absorbance is measured between 500-800 nm. However, the disadvantage of this assay is in its dependency for tyrosine residues. Also, it is highly sensitive to chemicals like TCA, Tris and EDTA. A much more sensitive assay involves the use of bicinchoninic acid after biuret reaction. The complex is even more stable than the same in Lowry assay and is highly sensitive for membrane proteomes. However, it is also affected by chemicals like EDTA, though at a higher concentration.

Majority of protein quantification methods for proteomic analysis employ Bradford's method, whereby a complex is formed between Coomassie brilliant blue and the proteins. Although the Coomassie dye prefers lysine and ariginine residues, the overall assay is highly sensitive and is compatible with many chemicals, unlike lowry or BCA assay. The dye on binding with the protein shows absorption maxima at 595 nm. However, at higher concentrations of proteins, there is a deviation from the linearity, which affects protein estimation to some extent. Nonetheless, Bradford assay is most commonly used for protein estimation in proteomic studies.

Challenges

Even though protein isolation is a routine work, it poses several challenges and these challenges arise depending on type of samples being handled. For example, algal and plant cells have a very strong cell wall, which is very difficult to disrupt by normal sonication method. Hence enzymatic pre-treatment or French press also accompanies sonication. Plant cells have high amount of pigments like chlorophyll and carotenoids and high amount of lipid. All these interfering compounds pose a challenge for preparation of good protein sample.

On the other hand, body fluids pose the problem of high salt contaminations and very dilute samples. Although both of these problems can be addressed simultaneously with the use of zip tipping (carbon columns), in certain cases like urine or cerebrospinal fluids, protein concentration is very low and hence isolation of proteins is extremely tricky. Also, the range of protein abundance varies in the body fluids and thus having a proteome profile of all the proteins becomes extremely difficult because many low abundant proteins get masked or are limited by the sensitivity of the technique used to identify them.

Bacterial cells are probably the easiest cells to harvest proteins, but due to their prokaryotic nature, the protein samples get highly contaminated with nucleic acids, which increase the viscosity of the medium and hence pose downstream problems. However, in all the samples, membrane proteins pose the greatest problem. They are highly hydrophobic in nature and hence are not readily soluble in the buffer generally used.

Also, they are extremely bound tightly with the membranes and disrupting the membranes at the level of individual building block is not possible. Hence, they are either under-represented or are totally lost from the proteome profile. Regardless of the sample to be analyzed, a combination of various methods is recommended for good protein sample preparation.

Protein Sample Preparation

A workflow of protein extraction from few important sample types is described below:

Bacterial

The bacterial cell is lysed by either gentle or harsh methods like lysozyme treatment and sonication, and the cell lysate is treated with TRIZol reagent (containing phenol and guanidium isothiocyanate) to denature the proteins. Chloroform and ethanol specifically separate out the RNA and the DNA. The protein is precipitated using acetone or a mixture of TCA and acetone. The advantage of TRIZol based method is that the protocol is devoid of any contamination from nucleic acid or lipids. There is also no need for desalting in this case. The proteins are highly soluble in the buffer and the yield is also very high.

Plant

Plant and algal samples being extremely tough are first crushed using liquid nitrogen and then the lysate is treated with either TCA/Acetone or by TRIZol reagent as described above. Plant cells contain pigments and hence a depigmentation step involving chloroform or hexane is required prior to protein extraction. This step also removes majority of the lipids, which can pose problem later if not removed.

Serum

Serum is one of the most important body fluid, which serves as the network of many important reactions. Hence, a study of the serum proteomics is very important. Nonetheless, serum proteomic analysis poses the challenges of high levels of protein abundance and salt contamination. Removal of these abundant proteins and interference results in enrichment of low-abundant proteins that can serve as potential biomarker for disease diagnosis.

References

- Fernández A, Scott R (2003). "Dehydron: a structurally encoded signal for protein interaction". Biophysical Journal. 85 (3): 1914–28. Bibcode:2003BpJ....85.1914F. PMC 1303363. PMID 12944304. doi:10.1016/S0006-3495(03)74619-0
- Dobson CM (2000). "The nature and significance of protein folding". In Pain RH. Mechanisms of Protein Folding. Oxford, Oxfordshire: Oxford University Press. pp. 1–28. ISBN 0-19-963789-X

- Wilkins, Marc (Dec 2009). "Proteomics data mining". Expert review of proteomics. England. 6 (6): 599–603. PMID 19929606. doi:10.1586/epr.09.81

- Zagrovic B, Snow CD, Shirts MR, Pande VS (2002). "Simulation of folding of a small alpha-helical protein in atomistic detail using worldwide-distributed computing". Journal of Molecular Biology. 323 (5): 927–37. PMID 12417204. doi:10.1016/S0022-2836(02)00997-X

- Sleator RD. (2012). "Prediction of protein functions". Methods in Molecular Biology. Methods in Molecular Biology. 815: 15–24. ISBN 978-1-61779-423-0. PMID 22130980. doi:10.1007/978-1-61779-424-7_2

- Gutteridge A, Thornton JM (2005). "Understanding nature's catalytic toolkit". Trends in Biochemical Sciences. 30 (11): 622–29. PMID 16214343. doi:10.1016/j.tibs.2005.09.006

- Hoffmann M, Wanko M, Strodel P, König PH, Frauenheim T, Schulten K, Thiel W, Tajkhorshid E, Elstner M (2006). "Color tuning in rhodopsins: the mechanism for the spectral shift between bacteriorhodopsin and sensory rhodopsin II". Journal of the American Chemical Society. 128 (33): 10808–18. PMID 16910676. doi:10.1021/ja062082i

- EBI External Services (2010-01-20). "The Catalytic Site Atlas at The European Bioinformatics Institute". Ebi.ac.uk. Retrieved 2011-01-16

- Hey J, Posch A, Cohen A, Liu N, Harbers A (2008). "Fractionation of complex protein mixtures by liquid-phase isoelectric focusing". Methods in Molecular Biology. Methods in Molecular Biology™. 424: 225–39. ISBN 978-1-58829-722-8. PMID 18369866. doi:10.1007/978-1-60327-064-9_19

- Kozlowski, Lukasz P. (2016). "Proteome-pI: proteome isoelectric point database". Nucleic Acids Research. 45 (D1): D1112–D1116. ISSN 0305-1048. PMC 5210655. PMID 27789699. doi:10.1093/nar/gkw978

- Ward, J. J.; Sodhi, J. S.; McGuffin, L. J.; Buxton, B. F.; Jones, D. T. (2004). "Prediction and functional analysis of native disorder in proteins from the three kingdoms of life". Journal of Molecular Biology. 337 (3): 635–45. PMID 15019783. doi:10.1016/j.jmb.2004.02.002

- Walker JH, Wilson K (2000). Principles and Techniques of Practical Biochemistry. Cambridge, UK: Cambridge University Press. pp. 287–89. ISBN 0-521-65873-X

- Fulton A, Isaacs W (1991). "Titin, a huge, elastic sarcomeric protein with a probable role in morphogenesis". BioEssays. 13 (4): 157–61. PMID 1859393. doi:10.1002/bies.950130403

- Brosnan J (June 2003). "Interorgan amino acid transport and its regulation". Journal of Nutrition. 133 (6 Suppl 1): 2068S–72S. PMID 12771367

- Peter Tompa; Alan Fersht (18 November 2009). Structure and Function of Intrinsically Disordered Proteins. CRC Press. ISBN 978-1-4200-7893-0

Protein Breaking and its Applications

Techniques are an important component of any field of study. The following section elucidates the various techniques that are related to protein-breaking. To study proteins, it is necessary to separate it from a mixture. Chromatography is a widely used method for separation of proteins. The method can be classified into affinity chromatography, adsorption chromatography, ion-exchange chromatography, partition chromatography, and molecular exclusion (gel filtration) chromatography.

Amino Acid

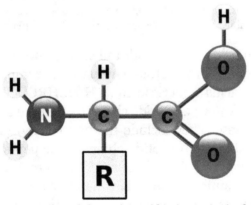

The structure of an alpha amino acid in its un-ionized form

Amino acids are organic compounds containing amine ($-NH_2$) and carboxyl (-COOH) functional groups, along with a side chain (R group) specific to each amino acid. The key elements of an amino acid are carbon, hydrogen, oxygen, and nitrogen, although other elements are found in the side chains of certain amino acids. About 500 amino acids are known (though only 20 appear in the genetic code) and can be classified in many ways. They can be classified according to the core structural functional groups' locations as alpha- (α-), beta- (β-), gamma- (γ-) or delta- (δ-) amino acids; other categories relate to polarity, pH level, and side chain group type (aliphatic, acyclic, aromatic, containing hydroxyl or sulfur, etc.). In the form of proteins, amino acid residues form the second-largest component (water is the largest) of human muscles and other tissues. Beyond their role as residues in proteins, amino acids participate in a number of processes such as neurotransmitter transport and biosynthesis.

The 21 proteinogenic α-amino acids found in eukaryotes, grouped according to their side chains' pK_a values and charges carried at physiological pH 7.4

In biochemistry, amino acids having both the amine and the carboxylic acid groups attached to the first (alpha-) carbon atom have particular importance. They are known as 2-, alpha-, or α-amino acids (generic formula $H_2NCHRCOOH$ in most cases, where R is an organic substituent known as a "side chain"); often the term "amino acid" is used to refer specifically to these. They include the 22 proteinogenic ("protein-building") amino acids, which combine into peptide chains ("polypeptides") to form the building-blocks of a vast array of proteins. These are all L-stereoisomers ("left-handed" isomers), although a few D-amino acids ("right-handed") occur in bacterial envelopes, as a neuromodulator (D-serine), and in some antibiotics. Twenty of the proteinogenic amino acids are encoded directly by triplet codons in the genetic code and are known as "standard" amino acids. The other two ("non-standard" or "non-canonical") are selenocysteine (present in many noneukaryotes as well as most eukaryotes, but not coded directly by DNA), and pyrrolysine (found only in some archea and one bacterium). Pyrrolysine and selenocysteine are encoded via variant codons; for example, selenocysteine is encoded by stop codon and SECIS element. *N*-formylmethionine (which is often the initial amino acid of proteins in bacteria, mitochondria, and chloroplasts) is generally considered as a form of methionine rather than as a separate proteinogenic amino acid. Codon–tRNA combinations not found in nature can also be used to "expand" the genetic code and create novel proteins known as alloproteins incorporating non-proteinogenic amino acids.

Many important proteinogenic and non-proteinogenic amino acids have biological functions. For example, in the human brain, glutamate (standard glutamic acid) and gamma-amino-butyric acid ("GABA", non-standard gamma-amino acid) are, respectively, the main excitatory and inhibitory neurotransmitters. Hydroxyproline, a major component of the connective tissue collagen), is synthesised from proline. Glycine is a biosynthetic precursor to porphyrins used in red blood cells. Carnitine is used in lipid transport.

Nine proteinogenic amino acids are called "essential" for humans because they cannot be created from other compounds by the human body and so must be taken in as food. Others may be conditionally essential for certain ages or medical conditions. Essential amino acids may also differ between species.

Because of their biological significance, amino acids are important in nutrition and are commonly used in nutritional supplements, fertilizers, and food technology. Industrial uses include the production of drugs, biodegradable plastics, and chiral catalysts.

History

The first few amino acids were discovered in the early 19th century. In 1806, French chemists Louis-Nicolas Vauquelin and Pierre Jean Robiquet isolated a compound in asparagus that was subsequently named asparagine, the first amino acid to be discovered. Cystine was discovered in 1810, although its monomer, cysteine, remained undiscovered until 1884. Glycine and leucine were discovered in 1820. The last of the 20 common amino acids to be discovered was threonine in 1935 by William Cumming Rose, who also determined the essential amino acids and established the minimum daily requirements of all amino acids for optimal growth.

Usage of the term *amino acid* in the English language is from 1898. Proteins were found to yield amino acids after enzymatic digestion or acid hydrolysis. In 1902, Emil Fischer and Franz Hofmeister proposed that proteins are the result of the formation of bonds between the amino group of one amino acid with the carboxyl group of another, in a linear structure that Fischer termed "peptide".

General Structure

In the structure shown at the top of the page, R represents a side chain specific to each amino acid. The carbon atom next to the carboxyl group (which is therefore numbered 2 in the carbon chain starting from that functional group) is called the α–carbon. Amino acids containing an amino group bonded directly to the alpha carbon are referred to as *alpha amino acids*. These include amino acids such as proline which contain secondary amines, which used to be often referred to as "imino acids".

Isomerism

The alpha amino acids are the most common form found in nature, but only when occurring in the L-isomer. The alpha carbon is a chiral carbon atom, with the exception of glycine which has two indistinguishable hydrogen atoms on the alpha carbon. Therefore, all alpha amino acids but glycine can exist in either of two enantiomers, called L or D amino acids, which are mirror images of each other. While L-amino acids represent all of the amino acids found in proteins during translation in the ribosome, D-amino acids are found in some proteins produced by enzyme posttranslational modifications after translation and translocation to the endoplasmic reticulum, as in exotic sea-dwelling organisms such as cone snails. They are also abundant components of the peptidoglycan cell walls of bacteria, and D-serine may act as a neurotransmitter in the brain. D-amino acids are used in racemic crystallography to create centrosymmetric crystals, which (depending on the protein) may allow for easier and more robust protein structure determination. The L and D convention for amino acid configuration refers not to the optical activity of the amino acid itself but rather to the optical activity of the isomer of glyceraldehyde from which that amino acid can, in theory, be synthesized (D-glyceraldehyde is dextrorotatory; L-glyceraldehyde is levorotatory). In alternative fashion, the *(S)* and *(R)* designators are used to indicate the absolute stereochemistry. Almost all of the amino acids in proteins are *(S)* at the α carbon, with cysteine being *(R)* and glycine non-chiral. Cysteine has its side chain in the same geometric position as the other amino acids, but the *R/S* terminology is reversed because of the higher atomic number of sulfur compared to the carboxyl oxygen gives the side chain a higher priority, whereas the atoms in most other side chains give them lower priority.

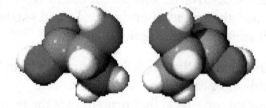

The two enantiomers of alanine, D-alanine and L-alanine

Side Chains

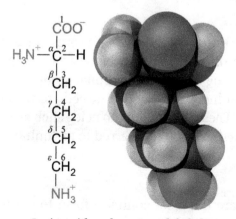

Lysine with carbon atoms labeled

In amino acids that have a carbon chain attached to the α–carbon the carbons are labeled in order as α, β, γ, δ, and so on. In some amino acids, the amine group is attached to the β or γ-carbon, and these are therefore referred to as *beta* or *gamma amino acids*.

Amino acids are usually classified by the properties of their side chain into four groups. The side chain can make an amino acid a weak acid or a weak base, and a hydrophile if the side chain is polar or a hydrophobe if it is nonpolar. The chemical structures of the 22 standard amino acids, along with their chemical properties.

The phrase "branched-chain amino acids" or BCAA refers to the amino acids having aliphatic side chains that are non-linear; these are leucine, isoleucine, and valine. Proline is the only proteinogenic amino acid whose side-group links to the α-amino group and, thus, is also the only proteinogenic amino acid containing a secondary amine at this position. In chemical terms, proline is, therefore, an imino acid, since it lacks a primary amino group, although it is still classed as an amino acid in the current biochemical nomenclature, and may also be called an "N-alkylated alpha-amino acid".

Zwitterions

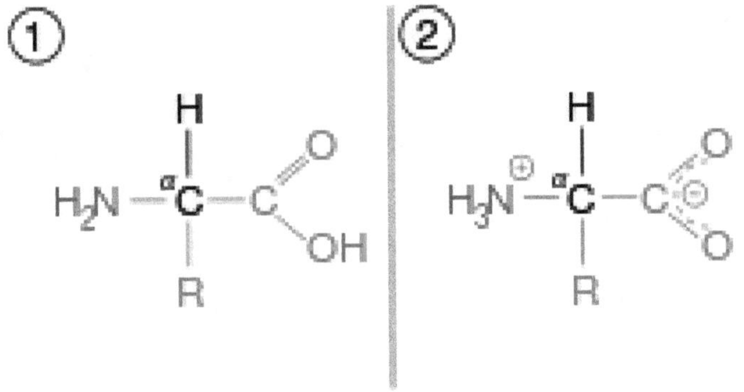

An amino acid in its (1) un-ionized and (2) zwitterionic forms

The α-carboxylic acid group of amino acids is a weak acid, meaning that it releases a hydron (such as a proton) at moderate pH values. In other words, carboxylic acid groups ($-CO_2H$) can be deprotonated to become negative carboxylates ($-CO_2^-$). The negatively charged carboxylate ion predominates at pH values greater than the pKa of the carboxylic acid group. In a complementary fashion, the α-amine of amino acids is a weak base, meaning that it accepts a proton at moderate pH values. In other words, α-amino groups (NH_2-) can be protonated to become positive α-ammonium groups ($^+NH_3-$). The positively charged α-ammonium group predominates at pH values less than the pKa of the α-ammonium group.

Because all amino acids contain amine and carboxylic acid functional groups, they share amphiprotic properties. Below pH 2.2, the predominant form will have a neutral carboxylic acid group and a positive α-ammonium ion (net charge +1), and above pH 9.4, a negative carboxylate and neutral α-amino group (net charge −1). But at pH between 2.2 and 9.4, an amino acid usually contains both a negative carboxylate and a positive α-ammonium group, as shown in structure (2), so has net zero charge. This molecular state is known as a zwitterion, from the German Zwitter meaning *hermaphrodite* or *hybrid*. The fully neutral form (structure (1)) is a very minor species in aqueous solution throughout the pH range (less than 1 part in 10^7). Amino acids exist as zwitterions also in the solid phase, and crystallize with salt-like properties unlike typical organic acids or amines.

Isoelectric Point

The variation in titration curves when the amino acids are grouped by category can be seen here. With the exception of tyrosine, using titration to differentiate between hydrophobic amino acids is problematic.

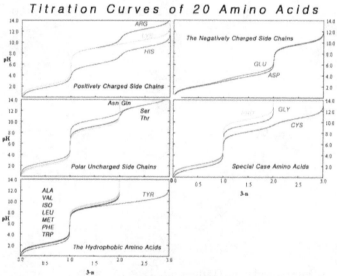

Composite of titration curves of twenty proteinogenic amino acids grouped by side chain category

At pH values between the two pKa values, the zwitterion predominates, but coexists in dynamic equilibrium with small amounts of net negative and net positive ions. At the exact midpoint between the two pKa values, the trace amount of net negative and trace of net positive ions exactly balance, so that average net charge of all forms present is zero. This pH is known as the isoelectric point pI, so pI = ½(pKa$_1$ + pKa$_2$). The individual amino acids all have slightly different pKa values, so have different isoelectric points. For amino acids with charged side chains, the pKa of the side chain is involved. Thus for Asp, Glu with negative side chains, pI = ½(pKa$_1$ + pKa$_R$), where pKa$_R$ is the side chain pKa. Cysteine also has potentially negative side chain with pKa$_R$ = 8.14, so pI should be calculated as for Asp and Glu, even though the side chain is not significantly charged at neutral pH. For His, Lys, and Arg with positive side chains, pI = ½(pKa$_R$

+ pKa_2). Amino acids have zero mobility in electrophoresis at their isoelectric point, although this behaviour is more usually exploited for peptides and proteins than single amino acids. Zwitterions have minimum solubility at their isoelectric point and some amino acids (in particular, with non-polar side chains) can be isolated by precipitation from water by adjusting the pH to the required isoelectric point.

Occurrence and Functions in Biochemistry

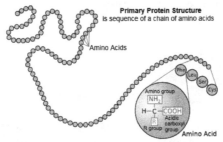

A polypeptide is an unbranched chain of amino acids

Proteinogenic Amino Acids

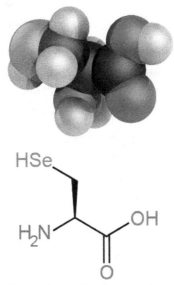

The amino acid selenocysteine

Amino acids are the structural units (monomers) that make up proteins. They join together to form short polymer chains called peptides or longer chains called either polypeptides or proteins. These polymers are linear and unbranched, with each amino acid within the chain attached to two neighboring amino acids. The process of making proteins is called *translation* and involves the step-by-step addition of amino acids to a growing protein chain by a ribozyme that is called a ribosome. The order in which the amino acids are added is read through the genetic code from an mRNA template, which is a RNA copy of one of the organism's genes.

Twenty-two amino acids are naturally incorporated into polypeptides and are called proteinogenic or natural amino acids. Of these, 20 are encoded by the universal genetic code. The remaining 2, selenocysteine and pyrrolysine, are incorporated into proteins by unique synthetic mechanisms. Selenocysteine is incorporated when the mRNA being translated includes a SECIS element, which causes the UGA codon to encode selenocysteine instead of a stop codon. Pyrrolysine is used by some methanogenic archaea in enzymes that they use to produce methane. It is coded for with the codon UAG, which is normally a stop codon in other organisms. This UAG codon is followed by a PYLIS downstream sequence.

Non-proteinogenic Amino Acids

L-α-alanine   β-alanine

β-alanine and its α-alanine isomer

Aside from the 22 proteinogenic amino acids, many *non-proteinogenic* amino acids are known. Those either are not found in proteins (for example carnitine, GABA, Levothyroxine) or are not produced directly and in isolation by standard cellular machinery (for example, hydroxyproline and selenomethionine).

Non-proteinogenic amino acids that are found in proteins are formed by post-translational modification, which is modification after translation during protein synthesis. These modifications are often essential for the function or regulation of a protein. Ffor example, the carboxylation of glutamate allows for better binding of calcium cations. connective tissues is composed of hydroxyproline, generated by hydroxylation of proline. Another example is the formation of hypusine in the translation initiation factor EIF5A, through modification of a lysine residue. Such modifications can also determine the localization of the protein, e.g., the addition of long hydrophobic groups can cause a protein to bind to a phospholipid membrane.

Some non-proteinogenic amino acids are not found in proteins. Examples include 2-aminoisobutyric acid and the neurotransmitter gamma-aminobutyric acid. Non-proteinogenic amino acids often occur as intermediates in the metabolic pathways for standard amino acids – for example, ornithine and citrulline occur in the urea cycle, part of amino acid catabolism. A rare exception to the dominance of α-amino acids in biology is the β-amino acid beta alanine (3-aminopropanoic acid), which is used in plants and microorganisms in the synthesis of pantothenic acid (vitamin B_5), a component of coenzyme A.

D-amino Acid Natural Abundance

D-isomers are uncommon in live organisms. For instance, gramicidin is a polypeptide made up from mixture of D- and L-amino acids. Other compounds containing D-amino acid are tyrocidine and valinomycin. These compounds disrupt bacterial cell walls, particularly in Gram-positive bacteria. Only 837 D-amino acids were found in Swiss-Prot database (187 million amino acids analysed).

Non-standard Amino Acids

The 20 amino acids that are encoded directly by the codons of the universal genetic code are called *standard* or *canonical* amino acids. A modified form of methionine (N-formylmethionine) is often incorporated in place of methionine as the initial amino acid of proteins in bacteria, mitochondria and chloroplasts. Other amino acids are called *non-standard* or *non-canonical*. Most of the non-standard amino acids are also non-proteinogenic (i.e. they cannot be incorporated into proteins during translation), but two of them are proteinogenic, as they can be incorporated translationally into proteins by exploiting information not encoded in the universal genetic code.

The two non-standard proteinogenic amino acids are selenocysteine (present in many non-eukaryotes as well as most eukaryotes, but not coded directly by DNA) and pyrrolysine (found only in some archaea and one bacterium). The incorporation of these non-standard amino acids is rare. For example, 25 human proteins include selenocysteine (Sec) in their primary structure, and the structurally characterized enzymes (selenoenzymes) employ Sec as the catalytic moiety in their active sites. Pyrrolysine and selenocysteine are encoded via variant codons. For example, selenocysteine is encoded by stop codon and SECIS element.

In Human Nutrition

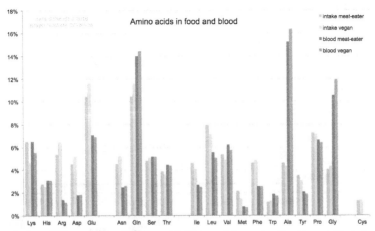

Share of amino acid in different human diets and the resulting mix of amino acids in human blood serum. Glutamate and glutamine are the most frequent in food at over 10%, while alanine, glutamine, and glycine are the most common in blood

When taken up into the human body from the diet, the 20 standard amino acids either are used to synthesize proteins and other biomolecules or are oxidized to urea and carbon dioxide as a source of energy. The oxidation pathway starts with the removal of the amino group by a transaminase; the amino group is then fed into the urea cycle. The other product of transamidation is a keto acid that enters the citric acid cycle. Glucogenic amino acids can also be converted into glucose, through gluconeogenesis. Of the 20 standard amino acids, nine (His, Ile, Leu, Lys, Met, Phe, Thr, Trp and Val), are called essential amino acids because the human body cannot synthesize them from other compounds at the level needed for normal growth, so they must be obtained from food. In addition, cysteine, taurine, tyrosine, and arginine are considered semiessential amino-acids in children (though taurine is not technically an amino acid), because the metabolic pathways that synthesize these amino acids are not fully developed. The amounts required also depend on the age and health of the individual, so it is hard to make general statements about the dietary requirement for some amino acids. Dietary exposure to the non-standard amino acid BMAA has been linked to human neurodegenerative diseases, including ALS.

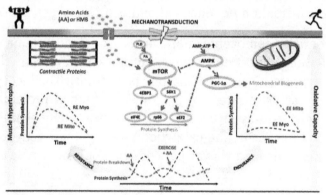

Diagram of the molecular signaling cascades that are involved in myofibrillar muscle protein synthesis and mitochondrial biogenesis in response to physical exercise and specific amino acids or their derivatives (primarily L-leucine and HMB). Many amino acids derived from food protein promote the activation of mTORC1 and increase protein synthesis by signaling through Rag GTPases

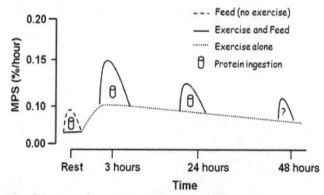

Resistance training stimulates muscle protein synthesis (MPS) for a period of up to 48 hours following exercise (shown by dotted line). Ingestion of a protein-rich meal at any point during this period will augment the exercise-induced increase in muscle protein synthesis (shown by solid lines)

Non-protein Functions

In humans, non-protein amino acids also have important roles as metabolic intermediates, such as in the biosynthesis of the neurotransmitter gamma-amino-butyric acid (GABA). Many amino acids are used to synthesize other molecules, for example:

- Tryptophan is a precursor of the neurotransmitter serotonin.

- Tyrosine (and its precursor phenylalanine) are precursors of the catecholamine neurotransmitters dopamine, epinephrine and norepinephrine and various trace amines.

- Phenylalanine is a precursor of phenethylamine and tyrosine in humans. In plants, it is a precursor of various phenylpropanoids, which are important in plant metabolism.

- Glycine is a precursor of porphyrins such as heme.

- Arginine is a precursor of nitric oxide.

- Ornithine and S-adenosylmethionine are precursors of polyamines.

- Aspartate, glycine, and glutamine are precursors of nucleotides. However, not all of the functions of other abundant non-standard amino acids are known.

Biosynthetic Pathways for Catecholamines and Trace Amines in the Human Brain

Catecholamines and trace amines are synthesized from phenylalanine and tyrosine in humans.

Some non-standard amino acids are used as defenses against herbivores in plants. For example, canavanine is an analogue of arginine that is found in many legumes, and in particularly large amounts in *Canavalia gladiata* (sword bean). This amino acid protects the plants from predators such as insects and can cause illness in people if some types of legumes are eaten without processing. The non-protein amino acid mimosine is found in other species of legume, in particular *Leucaena leucocephala*. This compound is an analogue of tyrosine and can poison animals that graze on these plants.

Uses in Industry

Amino acids are used for a variety of applications in industry, but their main use is as additives to animal feed. This is necessary, since many of the bulk components of these feeds, such as soybeans, either have low levels or lack some of the essential amino acids: lysine, methionine, threonine, and tryptophan are most important in the production of these feeds. In this industry, amino acids are also used to chelate metal cations in order to improve the absorption of minerals from supplements, which may be required to improve the health or production of these animals.

The food industry is also a major consumer of amino acids, in particular, glutamic acid, which is used as a flavor enhancer, and aspartame (aspartyl-phenyl-alanine-1-methyl ester) as a low-calorie artificial sweetener. Similar technology to that used for animal nutrition is employed in the human nutrition industry to alleviate symptoms of mineral deficiencies, such as anemia, by improving mineral absorption and reducing negative side effects from inorganic mineral supplementation.

The chelating ability of amino acids has been used in fertilizers for agriculture to facilitate the delivery of minerals to plants in order to correct mineral deficiencies, such as iron chlorosis. These fertilizers are also used to prevent deficiencies from occurring and improving the overall health of the plants. The remaining production of amino acids is used in the synthesis of drugs and cosmetics.

Similarly, some amino acids derivatives are used in pharmaceutical industry. They include 5-HTP (5-hydroxytryptophan) used for experimental treatment of depression, L-DOPA (L-dihydroxyphenylalanine) for Parkinson's treatment, and eflornithine drug that inhibits ornithine decarboxylase and used in the treatment of sleeping sickness.

Expanded Genetic Code

Since 2001, 40 non-natural amino acids have been added into protein by creating a unique codon (recoding) and a corresponding transfer-RNA:aminoacyl – tRNA-synthetase pair to encode it with diverse physicochemical and biological properties in order to be used as a tool to exploring protein structure and function or to create novel or enhanced proteins.

Nullomers

Nullomers are codons that in theory code for an amino acid, however in nature there is a selective bias against using this codon in favor of another, for example bacteria prefer to use CGA instead of AGA to code for arginine. This creates some sequences that do not appear in the genome. This characteristic can be taken advantage of and used to create new selective cancer-fighting drugs and to prevent cross-contamination of DNA samples from crime-scene investigations.

Chemical Building Blocks

Amino acids are important as low-cost feedstocks. These compounds are used in chiral pool synthesis as enantiomerically pure building-blocks.

Amino acids have been investigated as precursors chiral catalysts, e.g., for asymmetric hydrogenation reactions, although no commercial applications exist.

Biodegradable Plastics

Amino acids are under development as components of a range of biodegradable polymers. These materials have applications as environmentally friendly packaging and in medicine in drug delivery and the construction of prosthetic implants. These polymers include polypeptides, polyamides, polyesters, polysulfides, and polyurethanes with amino acids either forming part of their main chains or bonded as side chains. These modifications alter the physical properties and reactivities of the polymers. An interesting example of such materials is polyaspartate, a water-soluble biodegradable polymer that may have applications in disposable diapers and agriculture. Due to its solubility and ability to chelate metal ions, polyaspartate is also being used as a biodegradeable anti-scaling agent and a corrosion inhibitor. In addition, the aromatic amino acid tyrosine is being developed as a possible replacement for toxic phenols such as bisphenol A in the manufacture of polycarbonates.

Reactions

As amino acids have both a primary amine group and a primary carboxyl group, these chemicals can undergo most of the reactions associated with these functional groups. These include nucleophilic addition, amide bond formation, and imine formation for the amine group, and esterification, amide bond formation, and decarboxylation for the carboxylic acid group. The combination of these functional groups allow amino acids to be effective polydentate ligands for metal-amino acid chelates. The multiple side chains of amino acids can also undergo chemical reactions. The types of these reactions are determined by the groups on these side chains and are, therefore, different between the various types of amino acid.

The Strecker amino acid synthesis

Chemical Synthesis

Several methods exist to synthesize amino acids. One of the oldest methods begins with the bromination at the α-carbon of a carboxylic acid. Nucleophilic substitution with ammonia then converts the alkyl bromide to the amino acid. In alternative fashion, the Strecker amino acid synthesis involves the treatment of an aldehyde with potassium cyanide and ammonia, this produces an α-amino nitrile as an intermediate. Hydrolysis of the nitrile in acid then yields a α-amino acid. Using ammonia or ammonium salts in this reaction gives unsubstituted amino acids, whereas substituting primary and secondary amines will yield substituted amino acids. Likewise, using ketones, instead of aldehydes, gives α,α-disubstituted amino acids. The classical synthesis gives racemic mixtures of α-amino acids as products, but several alternative procedures using asymmetric auxiliaries or asymmetric catalysts have been developed.

At the current time, the most-adopted method is an automated synthesis on a solid support (e.g., polystyrene beads), using protecting groups (e.g., Fmoc and t-Boc) and activating groups (e.g., DCC and DIC).

Peptide Bond Formation

As both the amine and carboxylic acid groups of amino acids can react to form amide bonds, one amino acid molecule can react with another and become joined through an amide linkage. This polymerization of amino acids is what creates proteins. This condensation reaction yields the newly formed peptide bond and a molecule of water. In cells, this reaction does not occur directly; instead, the amino acid is first activated by attachment to a transfer RNA molecule through an ester bond. This aminoacyl-tRNA is produced in an ATP-dependent reaction carried out by an aminoacyl tRNA synthetase. This aminoacyl-tRNA is then a substrate for the ribosome, which catalyzes the attack of the amino group of the elongating protein chain on the ester bond. As a result of this mechanism, all proteins made by ribosomes are synthesized starting at their N-terminus and moving toward their C-terminus.

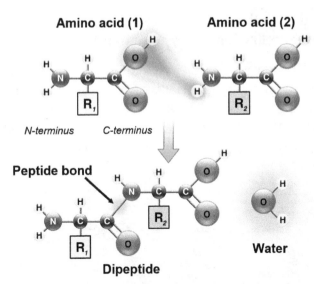

The condensation of two amino acids to form a *dipeptide* through a *peptide bond*

However, not all peptide bonds are formed in this way. In a few cases, peptides are synthesized by specific enzymes. For example, the tripeptide glutathione is an essential part of the defenses of cells against oxidative stress. This peptide is synthesized in two steps from free amino acids. In the first step, gamma-glutamylcysteine synthetase condenses cysteine and glutamic acid through a peptide bond formed between the side chain carboxyl of the glutamate (the gamma carbon of this side chain) and the amino group of the cysteine. This dipeptide is then condensed with glycine by glutathione synthetase to form glutathione.

In chemistry, peptides are synthesized by a variety of reactions. One of the most-used in solid-phase peptide synthesis uses the aromatic oxime derivatives of amino acids as activated units. These are added in sequence onto the growing peptide chain, which is attached to a solid resin support. The ability to easily synthesize vast numbers of different peptides by varying the types and order of amino acids (using combinatorial chemistry) has made peptide synthesis particularly important in creating libraries of peptides for use in drug discovery through high-throughput screening.

Biosynthesis

In plants, nitrogen is first assimilated into organic compounds in the form of glutamate, formed from alpha-ketoglutarate and ammonia in the mitochondrion. In order to form other amino acids, the plant uses transaminases to move the amino group to another alpha-keto carboxylic acid. For example, aspartate aminotransferase converts glutamate and oxaloacetate to alpha-ketoglutarate and aspartate. Other organisms use transaminases for amino acid synthesis, too.

Nonstandard amino acids are usually formed through modifications to standard amino acids. For example, homocysteine is formed through the transsulfuration pathway or by the demethylation of methionine via the intermediate metabolite S-adenosyl methionine, while hydroxyproline is made by a posttranslational modification of proline.

Microorganisms and plants can synthesize many uncommon amino acids. For example, some microbes make 2-aminoisobutyric acid and lanthionine, which is a sulfide-bridged derivative of alanine. Both of these amino acids are found in peptidic lantibiotics such as alamethicin. However, in plants, 1-aminocyclopropane-1-carboxylic acid is a small disubstituted cyclic amino acid that is a key intermediate in the production of the plant hormone ethylene.

Catabolism

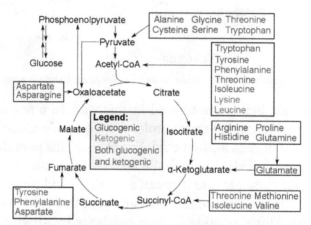

Catabolism of proteinogenic amino acids. Amino acids can be classified according to the properties of their main products as either of the following:
* *Glucogenic*, with the products having the ability to form glucose by gluconeogenesis
* *Ketogenic*, with the products not having the ability to form glucose. These products may still be used for ketogenesis or lipid synthesis.
* Amino acids catabolized into both glucogenic and ketogenic products.

Amino acids must first pass out of organelles and cells into blood circulation via amino acid transporters, since the amine and carboxylic acid groups are typically ionized. Degradation of an amino acid, occurring in the liver and kidneys, often involves deamination by moving its amino group to alpha-ketoglutarate, forming glutamate. This process involves transaminases, often the same as those used in amination during synthesis. In many vertebrates, the amino group is then removed through the urea cycle and is excreted in the form of urea. However, amino acid degradation can produce uric acid or ammonia instead. For example, serine dehydratase converts serine to pyruvate and ammonia. After removal of one or more amino groups, the remainder of the molecule can sometimes be used to synthesize new amino acids, or it can be used for energy by entering glycolysis or the citric acid cycle.

Physicochemical Properties of Amino Acids

The 20 amino acids encoded directly by the genetic code can be divided into several groups based on their properties. Important factors are charge, hydrophilicity or hydrophobicity, size, and functional groups. These properties are important for protein structure and protein–protein interactions. The water-soluble proteins tend to have their hydrophobic residues (Leu, Ile, Val, Phe, and Trp) buried in the middle of the protein, whereas hydrophilic side chains are exposed to the aqueous solvent. (Note that in biochemistry, a residue refers to a specific monomer within the polymeric chain of a polysaccharide, protein or nucleic acid.) The integral membrane proteins tend to have outer rings of exposed hydrophobic amino acids that anchor them into the lipid bilayer. In the case part-way between these two extremes, some peripheral membrane proteins have a patch of hydrophobic amino acids on their surface that locks onto the membrane. In similar fashion, proteins that have to bind to positively charged molecules have surfaces rich with negatively charged amino acids like glutamate and aspartate, while proteins binding to negatively charged molecules have surfaces rich with positively charged chains like lysine and arginine. There are different hydrophobicity scales of amino acid residues.

Some amino acids have special properties such as cysteine, that can form covalent disulfide bonds to other cysteine residues, proline that forms a cycle to the polypeptide backbone, and glycine that is more flexible than other amino acids.

Many proteins undergo a range of posttranslational modifications, when additional chemical groups are attached to the amino acids in proteins. Some modifications can produce hydrophobic lipoproteins, or hydrophilic glycoproteins. These type of modification allow the reversible targeting of a protein to a membrane. For example, the addition and removal of the fatty acid palmitic acid to cysteine residues in some signaling proteins causes the proteins to attach and then detach from cell membranes.

Table of Standard Amino Acid Abbreviations and Properties

Amino Acid	3-Letter	1-Letter	Side chain class	Side chain polarity	Side chain charge (pH 7.4)	Hydropathy index	Absorbance λ_{max}(nm)	ε at λ_{max} (mM^{-1} cm^{-1})	MW (Weight)	Occurrence in proteins (%)
Alanine	Ala	A	aliphatic	nonpolar	neutral	1.8			89.094	8.76
Arginine	Arg	R	basic	basic polar	positive	−4.5			174.203	5.78
Asparagine	Asn	N	amide	polar	neutral	−3.5			132.119	3.93
Aspartic acid	Asp	D	acid	acidic polar	negative	−3.5			133.104	5.49

Cysteine	Cys	C	sulfur-containing	nonpolar	neutral	2.5	250	0.3	121.154	1.38
Glutamic acid	Glu	E	acid	acidic polar	negative	−3.5			147.131	6.32
Glutamine	Gln	Q	amide	polar	neutral	−3.5			146.146	3.9
Glycine	Gly	G	aliphatic	nonpolar	neutral	−0.4			75.067	7.03
Histidine	His	H	basic aromatic	basic polar	positive(10%) neutral(90%)	−3.2	211	5.9	155.156	2.26
Isoleucine	Ile	I	aliphatic	nonpolar	neutral	4.5			131.175	5.49
Leucine	Leu	L	aliphatic	nonpolar	neutral	3.8			131.175	9.68
Lysine	Lys	K	basic	basic polar	positive	−3.9			146.189	5.19
Methionine	Met	M	sulfur-containing	nonpolar	neutral	1.9			149.208	2.32
Phenylalanine	Phe	F	aromatic	nonpolar	neutral	2.8	257, 206, 188	0.2, 9.3, 60.0	165.192	3.87
Proline	Pro	P	cyclic	nonpolar	neutral	−1.6			115.132	5.02
Serine	Ser	S	hydroxyl-containing	polar	neutral	−0.8			105.093	7.14
Threonine	Thr	T	hydroxyl-containing	polar	neutral	−0.7			119.119	5.53
Tryptophan	Trp	W	aromatic	nonpolar	neutral	−0.9	280, 219	5.6, 47.0	204.228	1.25
Tyrosine	Tyr	Y	aromatic	polar	neutral	−1.3	274, 222, 193	1.4, 8.0, 48.0	181.191	2.91
Valine	Val	V	aliphatic	nonpolar	neutral	4.2			117.148	6.73

Two additional amino acids are in some species coded for by codons that are usually interpreted as stop codons:

21st and 22nd amino acids	3-Letter	1-Letter	MW(Weight)
Selenocysteine	Sec	U	168.064
Pyrrolysine	Pyl	O	255.313

In addition to the specific amino acid codes, placeholders are used in cases where chemical or crystallographic analysis of a peptide or protein cannot conclusively determine the identity of a residue. They are also used to summarise conserved protein sequence motifs. The use of single letters to indicate sets of similar residues is similar to the use of abbreviation codes for degenerate bases.

Ambiguous Amino Acids	3-Letter	1-Letter	
Any / unknown	Xaa	X	All
Asparagine or aspartic acid	Asx	B	D, N
Glutamine or glutamic acid	Glx	Z	E, Q
Leucine or Isoleucine	Xle	J	I, L
Hydrophobic		Φ	V, I, L, F, W, Y, M
Aromatic		Ω	F, W, Y, H
Aliphatic (Non-Aromatic)		Ψ	V, I, L, M
Small		π	P, G, A, S
Hydrophilic		ζ	S, T, H, N, Q, E, D, K, R
Positively charged		+	K, R, H
Negatively charged		−	D, E

Unk is sometimes used instead of Xaa, but is less standard.

In addition, many non-standard amino acids have a specific code. For example, several peptide drugs, such as Bortezomib and MG132, are artificially synthesized and retain their protecting groups, which have specific codes. Bortezomib is Pyz-Phe-boroLeu, and MG132 is Z-Leu-Leu-Leu-al. To aid in the analysis of protein structure, photo-reactive amino acid analogs are available. These include photoleucine (pLeu) and photomethionine (pMet).

Peptide Bond

The covalent bond that holds two adjacent amino acid residues together is known as a peptide bond, which is formed between the carboxyl group of one amino acid and the amino group of other amino acid and is accompanied by release of a water molecule. The peptide bond is stabilized by resonance structure. The amide bond exhibits partial double bond character and is planar. In other words, it can exist in "cis" and "trans" form. In the unfolded form of a given protein, the peptide bonds have the liberty to take up either of the two forms; however the folded conformation has the peptide bond in a single form alone. The "trans" form is usually preferred as it's conformation is stable as compared to the "cis" form (Exception: Proline, which can exist in "cis" as well as "trans" form). The psi and phi are the angles of rotation about the bond between the a-carbon atom and carboxyl and amino groups, respectively. These angles determine which protein conformations will be favourable during protein folding.

Protein Structure

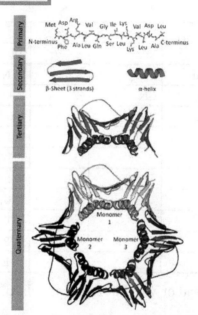

Protein structure, using PCNA as an example. (PDB: 1AXC)

Protein structure is the three-dimensional arrangement of atoms in a protein molecule. Proteins are polymers — specifically polypeptides — formed from sequences of amino acids, the monomers of the polymer. A single amino acid monomer may also be called a residue indicating a repeating unit of a polymer. Proteins form by amino acids undergoing condensation reactions, in which the amino acids lose one water molecule per reaction in order to attach to one another with a peptide bond. By convention, a chain under 30 amino acids is often identified as a peptide, rather than a protein. To be able to perform their biological function, proteins fold into one or more specific spatial conformations driven by a number of non-covalent interactions such as hydrogen bonding, ionic interactions, Van der Waals forces, and hydrophobic packing. To understand the functions of proteins at a molecular level, it is often necessary to determine their three-dimensional structure. This is the topic of the scientific field of structural biology, which employs techniques such as X-ray crystallography, NMR spectroscopy, and dual polarisation interferometry to determine the structure of proteins.

Protein structures range in size from tens to several thousand amino acids. By physical size, proteins are classified as nanoparticles, between 1–100 nm. Very large aggregates can be formed from protein subunits. For example, many thousands of actin molecules assemble into a microfilament.

A protein may undergo reversible structural changes in performing its biological function. The alternative structures of the same protein are referred to as different conformational isomers, or simply, conformations, and transitions between them are called conformational changes.

Levels of Protein Structure

There are four distinct levels of protein structure.

Amino Acid Residues

Each α-amino acid consists of a backbone that is present in all the amino acid types and a side chain that is unique to each type of residue. An exception from this rule is proline. Because the carbon atom is bound to four different groups it is chiral, however only one of the isomers occur in biological proteins. Glycine however, is not chiral since its side chain is a hydrogen atom. A simple mnemonic for correct L-form is "CORN": when the C_α atom is viewed with the H in front, the residues read "CO-R-N" in a clockwise direction.

Primary Structure

The primary structure of a protein refers to the linear sequence of amino acids in the polypeptide chain. The primary structure is held together by covalent bonds such as peptide bonds, which are made during the process of protein biosynthesis. The two ends of the polypeptide chain are referred to as the carboxyl terminus (C-terminus) and the amino terminus (N-terminus) based on the nature of the free group on each extremity. Counting of residues always starts at the N-terminal end (NH_2-group), which is the end where the amino group is not involved in a peptide bond. The primary structure of a protein is determined by the gene corresponding to the protein. A specific sequence of nucleotides in DNA is transcribed into mRNA, which is read by the ribosome in a process called translation. The sequence of amino acids in insulin was discovered by Frederick Sanger, establishing that proteins have defining amino acid sequences. The sequence of a protein is unique to that protein, and defines the structure and function of the protein. The sequence of a protein can be determined by methods such as Edman degradation or tandem mass spectrometry. Often, however, it is read directly from the sequence of the gene using the genetic code. It is strictly recommended to use the words "amino acid residues" when discussing proteins because when a peptide bond is formed, a water molecule is lost, and therefore proteins are made up of amino acid residues. Post-translational modification such as disulfide bond formation, phosphorylations and glycosylations are usually also considered a part of the primary structure, and cannot be read from the gene. For example, insulin is composed of 51 amino acids in 2 chains. One chain has 31 amino acids, and the other has 20 amino acids.

Secondary Structure

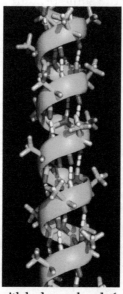

An α-helix with hydrogen bonds (yellow dots)

Secondary structure refers to highly regular local sub-structures on the actual polypeptide backbone chain. Two main types of secondary structure, the α-helix and the β-strand or β-sheets, were suggested in 1951 by Linus Pauling and coworkers. These secondary structures are defined by patterns of hydrogen bonds between the main-chain peptide groups. They have a regular geometry, being constrained to specific values of the dihedral angles ψ and φ on the Ramachandran plot. Both the α-helix and the β-sheet represent a way of saturating all the hydrogen bond donors and acceptors in the peptide backbone. Some parts of the protein are ordered but do not form any regular structures. They should not be confused with random coil, an unfolded polypeptide chain lacking any fixed three-dimensional structure. Several sequential secondary structures may form a "supersecondary unit".

Tertiary Structure

Tertiary structure refers to the three-dimensional structure of monomeric and multimeric protein molecules. The α-helixes and β-pleated-sheets are folded into a compact globular structure. The folding is driven by the *non-specific* hydrophobic interactions, the burial of hydrophobic residues from water, but the structure is stable only when the parts of a protein domain are locked into place by *specific* tertiary interactions, such as salt bridges, hydrogen bonds, and the tight packing of side chains and disulfide bonds. The disulfide bonds are extremely rare in cytosolic proteins, since the cytosol (intracellular fluid) is generally a reducing environment.

Quaternary Structure

Quaternary structure is the three-dimensional structure of a multi-subunit protein and how the subunits fit together. In this context, the quaternary structure is stabilized by the same non-covalent interactions and disulfide bonds as the tertiary structure. Complexes of two or more polypeptides (i.e. multiple subunits) are called multimers. Specifically it would be called a dimer if it contains two subunits, a trimer if it contains three subunits, a tetramer if it contains four subunits, and a pentamer if it contains five subunits. The subunits are frequently related to one another by symmetry operations, such as a 2-fold axis in a dimer. Multimers made up of identical subunits are referred to with a prefix of "homo-" (e.g. a homotetramer) and those made up of different subunits are referred to with a prefix of "hetero-", for example, a heterotetramer, such as the two alpha and two beta chains of hemoglobin.

Domains, Motifs, and Folds in Protein Structure

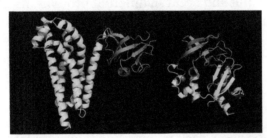

Protein domains. The two shown protein structures share a common domain (maroon), the PH domain, which is involved in phosphatidylinositol (3,4,5)-trisphosphate binding

Proteins are frequently described as consisting of several structural units. These units include domains, motifs, and folds. Despite the fact that there are about 100,000 different proteins expressed in eukaryotic systems, there are many fewer different domains, structural motifs and folds.

Structural Domain

A structural domain is an element of the protein's overall structure that is self-stabilizing and often folds independently of the rest of the protein chain. Many domains are not unique to the protein products of one gene or one gene family but instead appear in a variety of proteins. Domains often are named and singled out because they figure prominently in the biological function of the protein they belong to; for example, the "calcium-binding domain of calmodulin". Because they are independently stable, domains can be "swapped" by genetic engineering between one protein and another to make chimera proteins.

Structural and Sequence Motif

The structural and sequence motifs refer to short segments of protein three-dimensional structure or amino acid sequence that were found in a large number of different proteins.

Supersecondary Structure

The supersecondary structure refers to a specific combination of secondary structure elements, such as β-α-β units or a helix-turn-helix motif. Some of them may be also referred to as structural motifs.

Protein Fold

A protein fold refers to the general protein architecture, like a helix bundle, β-barrel, Rossman fold or different "folds" provided in the Structural Classification of Proteins database. A related concept is protein topology that refers to the arrangement of contacts within the protein.

Superdomain

A superdomain consists of two or more nominally unrelated structural domains that are inherited as a single unit and occur in different proteins. An example is provided by the protein tyrosine phosphatase domain and C2 domain pair in PTEN, several tensin proteins, auxilin and proteins in plants and fungi. The PTP-C2 superdomain evidently came into existence prior to the divergence of fungi, plants and animals is therefore likely to be about 1.5 billion years old.

Protein Folding

Once translated by a ribosome, each polypeptide folds into its characteristic three-dimensional structure from a random coil. Since the fold is maintained by a network of interactions between amino acids in the polypeptide, the native state of the protein chain is determined by the amino acid sequence (Anfinsen's dogma).

Protein Structure Determination

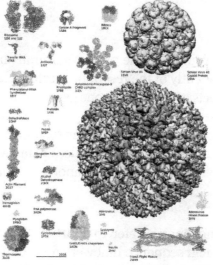

Examples of protein structures from the PDB

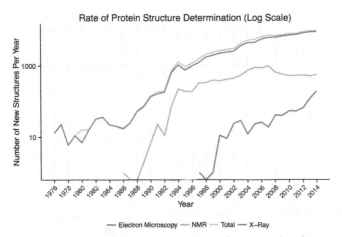

Rate of Protein Structure Determination by Method and Year

Around 90% of the protein structures available in the Protein Data Bank have been determined by X-ray crystallography. This method allows one to measure the three-dimensional (3-D) density distribution of electrons in the protein, in the crystallized state, and thereby infer the 3-D coordinates of all the atoms to be determined to a certain resolution. Roughly 9% of the known protein structures have been obtained by nuclear magnetic resonance techniques. The secondary structure composition can be determined via circular dichroism. Vibrational spectroscopy can also be used to characterize the conformation of peptides, polypeptides, and proteins. Two-dimensional infrared spectroscopy has become a valuable method to investigate the structures of flexible peptides and proteins that cannot be studied with other methods. Cryo-electron microscopy has recently become a means of determining protein structures to high resolution, less than 5 ångströms or 0.5 nanometer, and is anticipated to increase in power as a tool for high resolution work in the next decade. This technique is still a valuable resource for researchers working with very large protein complexes such as virus coat proteins and amyloid fibers. A more qualitative picture of protein structure is often obtained by proteolysis, which is also useful to screen for more crystallizable protein samples. Novel implementations of this approach, including fast parallel proteolysis (FASTpp), can probe the structured fraction and its stability without the need for purification.

Protein Sequence Analysis: Ensembles

Proteins are often thought of as relatively stable structures that have a set tertiary structure and experience conformational changes as a result of being modified by other proteins or as part of enzymatic activity. However proteins have varying degrees of stability and some of the less stable variants are intrinsically disordered proteins. These proteins exist and function in a relatively 'disordered' state lacking a stable tertiary structure. As a result, they are difficult to describe in a standard protein structure model that was designed for proteins with a fixed tertiary structure. Conformational ensembles have been devised as a way to provide a more accurate and 'dynamic' representation of the conformational state of intrinsically disordered proteins. Conformational ensembles function by attempting to represent the various conformations of intrinsically disordered proteins within an ensemble file (the type found at the Protein Ensemble Database).

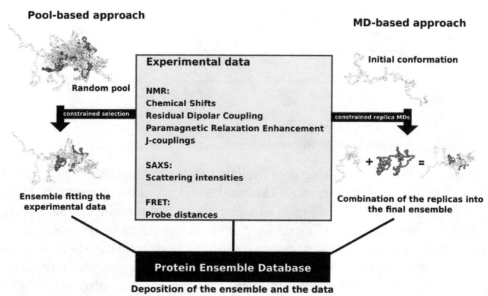

Schematic view of the two main ensemble modeling approaches

Protein ensemble files are a representation of a protein that can be considered to have a flexible structure. Creating these files requires determining which of the various theoretically possible protein conformations actually exist. One approach is to apply computational algorithms to the protein data in order to try to determine the most likely set of conformations for an ensemble file.

There are multiple methods for preparing data for the Protein Ensemble Database that fall into two general methodologies – pool and molecular dynamics (MD) approaches (diagrammed in the figure). The pool based approach uses the protein's amino acid sequence to create a massive pool of random conformations. This pool is then subjected to more computational processing that creates a set of theoretical parameters for each conformation based on the structure. Conformational subsets from this pool whose average theoretical parameters closely match known experimental data for this protein are selected.

The molecular dynamics approach takes multiple random conformations at a time and subjects all of them to experimental data. Here the experimental data is serving as limitations to be placed on the conformations (e.g. known distances between atoms). Only conformations that manage to remain within the limits set by the experimental data are accepted. This approach often applies large amounts of experimental data to the conformations which is a very computationally demanding task.

Protein	Data Type	Protocol	PED ID
Sic1/Cdc4	NMR and SAXS	Pool-based	PED9AAA
p15 PAF	NMR and SAXS	Pool-based	PED6AAA
MKK7	NMR	Pool-based	PED5AAB
Beta-synuclein	NMR	MD-based	PED1AAD
P27 KID	NMR	MD-based	PED2AAA

Structure Classification

Protein structures can be grouped based on their similarity or a common evolutionary origin. The Structural Classification of Proteins database and CATH database provide two different structural classifications of proteins. Shared structure between proteins is considered evidence of evolutionary relatedness between proteins and is used group proteins together into protein superfamilies.

Computational Prediction of Protein Structure

The generation of a protein sequence is much easier than the determination of a protein structure. However, the structure of a protein gives much more insight in the function of the protein than its sequence. Therefore, a number of methods for the computational prediction of protein structure from its sequence have been developed. *Ab initio* prediction methods use just the sequence of the protein. Threading and homology modeling methods can build a 3-D model for a protein of unknown structure from experimental structures of evolutionarily-related proteins, called a protein family.

Protein Folding

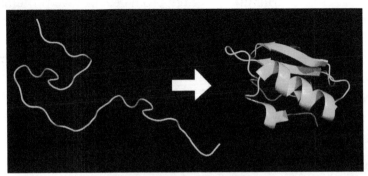

Protein before and after folding

Protein folding is the physical process by which a protein chain acquires its native 3-dimensional structure, a conformation that is usually biologically functional, in an expeditious and reproducible manner. It is the physical process by which a polypeptide folds into its characteristic and functional three-dimensional structure from random coil. Each protein exists as an unfolded polypeptide or random coil when translated from a sequence of mRNA to a linear chain of amino acids. This polypeptide lacks any stable (long-lasting) three-dimensional structure (the left hand side of the first figure). As the polypeptide chain is being synthesized by the ribosome, the linear chain begins to fold into its three dimensional structure. Folding begins to occur even during translation of the polypeptide chain. Amino acids interact with each other to produce a well-defined three-dimensional structure, the folded protein, known as the native state. The resulting three-dimensional structure is determined by the amino acid sequence or primary structure (Anfinsen's dogma). The energy landscape describes the folding pathways in which the unfolded protein is able to assume its native state. Experiments beginning in the 1980s indicate the codon for an amino acid can also influence protein structure.

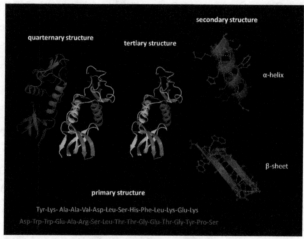

Results of protein folding

The correct three-dimensional structure is essential to function, although some parts of functional proteins may remain unfolded, so that protein dynamics is important. Failure to fold into native structure generally produces inactive proteins, but in some instances misfolded proteins have modified or toxic functionality. Several neurodegenerative and other diseases are believed to result from the accumulation of amyloid fibrils formed by *misfolded* proteins. Many allergies are caused by incorrect folding of some proteins, because the immune system does not produce antibodies for certain protein structures.

Process of Protein Folding

Primary Structure

The primary structure of a protein, its linear amino-acid sequence, determines its native conformation. The specific amino acid residues and their position in the polypeptide chain are the determining factors for which portions of the protein fold closely together and form its three dimensional conformation. The amino acid composition is not as important as the sequence. The essential fact of folding, however, remains that the amino acid sequence of each protein contains the information that specifies both the native structure and the pathway to attain that state. This is not to say that nearly identical amino acid sequences always fold similarly. Conformations differ based on environmental factors as well; similar proteins fold differently based on where they are found.

Secondary Structure

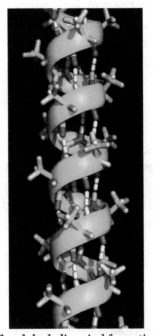

The alpha helix spiral formation

An anti-parallel beta pleated sheet displaying hydrogen bonding within the backbone

Formation of a secondary structure is the first step in the folding process that a protein takes to assume its native structure. Characteristic of secondary structure are the structures known as alpha helices and beta sheets that fold rapidly because they are stabilized by intramolecular hydrogen bonds, as was first characterized by Linus Pauling. Formation of intramolecular hydrogen bonds provides another important contribution to protein stability. Alpha helices are formed by hydrogen bonding of the backbone to form a spiral shape. The beta pleated sheet is a structure that forms with the backbone bending over itself to form the hydrogen bonds. The hydrogen bonds are between the amide hydrogen and carbonyl carbon of the peptide bond. There exists anti-parallel beta pleated sheets and parallel beta pleated sheets where the stability of the hydrogen bonds is stronger in the anti-parallel beta sheet as it hydrogen bonds with the ideal 180 degree angle compared to the slanted hydrogen bonds formed by parallel sheets. The alpha helices and beta pleated sheets can be amphipathic in nature, or contain a hydrophilic portion and a hydrophobic portion. This property of these secondary structures aids in the folding of the protein as it aligns the helices and sheets in such a way where the hydrophilic sides are facing the aqueous environment surrounding the protein and the hydrophobic sides are facing the hydrophobic core of the protein. Secondary structure hierarchically gives way to tertiary structure formation.

Tertiary Structure

Once the protein's tertiary structure is formed and stabilized by the hydrophobic interactions, there may also be covalent bonding in the form of disulfide bridges formed between two cysteine residues. Tertiary structure of a protein involves a single polypeptide chain; however, additional interactions of folded polypeptide chains give rise to quaternary structure formation.

Quaternary Structure

Tertiary structure may give way to the formation of quaternary structure in some proteins, which usually involves the "assembly" or "coassembly" of subunits that have already folded; in other words, multiple polypeptide chains could interact to form a fully functional quaternary protein.

Driving Force of Protein Folding

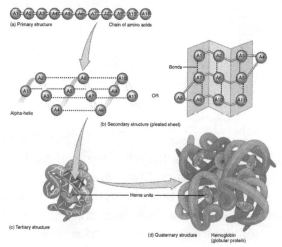

All forms of protein structure summarized

Folding is a spontaneous process that is mainly guided by hydrophobic interactions, formation of intramolecular hydrogen bonds, van der Waals forces, and it is opposed by conformational entropy. The process of folding often begins co-translationally, so that the N-terminus of the protein begins to fold while the C-terminal portion of the protein is still being synthesized by the ribosome; however, a protein molecule may fold spontaneously during or after biosynthesis. While these macromolecules may be regarded as "folding themselves", the process also depends on the solvent (water or lipid bilayer), the concentration of salts, the pH, the temperature, the possible presence of cofactors and of molecular chaperones. Proteins will have limitations on their folding abilities by the restricted bending angles or conformations that are possible. These allowable angles of protein folding are described with a two-dimensional plot known as the Ramachandran plot, depicted with psi and phi angles of allowable rotation.

Hydrophobic Effect

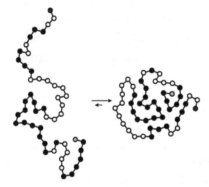

Hydrophobic collapse. In the compact fold (to the right), the hydrophobic amino acids (shown as black spheres) collapse toward the center to become shielded from aqueous environment

Protein folding must be thermodynamically favorable within a cell in order for it to be a spontaneous reaction. Since it is known that protein folding is a spontaneous reaction, then it must assume a negative Gibbs free energy value. Gibbs free energy in protein folding is directly related to enthalpy and entropy. For a negative delta G to arise and for protein folding to become thermodynamically favorable, then either enthalpy, entropy, or both terms must be favorable.

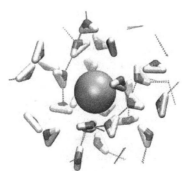

Entropy is decreased as the water molecules become more orderly near the hydrophobic solute

Minimizing the number of hydrophobic side-chains exposed to water is an important driving force behind the folding process. The hydrophobic effect is the phenomenon in which the hydrophobic chains of a protein collapse into the core of the protein (away from the hydrophilic environment). In an aqueous environment, the water molecules tend to aggregate around the hydrophobic regions or side chains of the protein, creating water shells of ordered water molecules. An ordering of water molecules around a hydrophobic region increases order in a system and therefore contributes a negative change in entropy (less entropy in the system). The water molecules are fixed in these water cages which drives the hydrophobic collapse, or the inward folding of the hydrophobic groups. The hydrophobic collapse introduces entropy back to the system via the breaking of the water cages which frees the ordered water molecules. The multitude of hydrophobic groups interacting within the core of the globular folded protein contributes a significant amount to protein stability after folding, because of the vastly accumulated van der Waals forces (specifically London Dispersion forces). The hydrophobic effect exists as a driving force in thermodynamics only if there is the presence of an aqueous medium with an amphiphilic molecule containing a large hydrophobic region. The strength of hydrogen bonds depends on their environment; thus, H-bonds enveloped in a hydrophobic core contribute more than H-bonds exposed to the aqueous environment to the stability of the native state.

Chaperones

Chaperones are a class of proteins that aid in the correct folding of other proteins *in vivo*. Chaperones exist in all cellular compartments and interact with the polypeptide chain in order to allow the native three-dimensional conformation of the protein to form; however, chaperones themselves are not included in the final structure of the protein they are assisting in. Chaperones may assist in folding even when the nascent polypeptide is being synthesized by the ribosome. Molecular chaperones operate by binding to stabilize an otherwise unstable structure of a protein in its folding pathway, but chaperones do not contain the necessary information to know the correct native structure of the protein they are aiding; rather, chaperones work by preventing incorrect folding conformations. In this way, chaperones do not actually increase the rate of individual steps involved in the folding pathway toward the native structure; instead, they work by reducing possible unwanted aggregations of the polypeptide chain that might otherwise slow down the search for the proper intermediate and they provide a more efficient pathway for the polypeptide chain to assume the correct conformations. Chaperones are not to be confused with folding catalysts, which actually do catalyze the otherwise slow steps in the folding pathway. Examples of folding catalysts are protein disulfide isomerases and peptidyl-prolyl isomerases that may be involved in formation of disulfide bonds or interconversion between cis and trans stereoisomers, respectively. Chaperones are shown to be critical in the process of protein folding *in vivo* because they provide the protein with the aid needed to assume its proper alignments and conformations efficiently enough to become "biologically relevant". This means that the polypeptide chain could theoretically fold into its native structure without the aid of chaperones, as demonstrated by protein folding experiments conducted *in vitro*; however, this process proves to be too inefficient or too slow to exist in biological systems; therefore, chaperones are necessary for protein folding *in vivo*. Along with its role in aiding native structure formation, chaperones are shown to be involved in various roles such as protein transport, degradation, and even allow denatured proteins exposed to certain external denaturant factors an opportunity to refold into their correct native structures.

Example of a small eukaryotic heat shock protein

A fully denatured protein lacks both tertiary and secondary structure, and exists as a so-called random coil. Under certain conditions some proteins can refold; however, in many cases, denaturation is irreversible. Cells sometimes protect their proteins against the denaturing influence of heat with enzymes known as heat shock proteins (a type of chaperone), which assist other proteins both in folding and in remaining folded. Some proteins never fold in cells at all except with the assistance of chaperones which either isolate individual proteins so that their folding is not interrupted by interactions with other proteins or help to unfold misfolded proteins, allowing them to refold into the correct native structure. This function is crucial to prevent the risk of precipitation into insoluble amorphous aggregates. The external factors involved in protein denaturation or disruption of the native state include temperature, external fields (electric, magnetic), molecular crowding, and even the limitation of space, which can have a big influence on the folding of proteins. High concentrations of solutes, extremes of pH, mechanical forces, and the presence of chemical denaturants can contribute to protein denaturation, as well. These individual factors are categorized together as stresses. Chaperones are shown to exist in increasing concentrations during times of cellular stress and help the proper folding of emerging proteins as well as denatured or misfolded ones.

Under some conditions proteins will not fold into their biochemically functional forms. Temperatures above or below the range that cells tend to live in will cause thermally unstable proteins to unfold or denature (this is why boiling makes an egg white turn opaque). Protein thermal stability is far from constant, however; for example, hyperthermophilic bacteria have been found that grow at temperatures as high as 122 °C, which of course requires that their full complement of vital proteins and protein assemblies be stable at that temperature or above.

Computational Methods for Studying Protein Folding

The study of protein folding includes three main aspects related to the prediction of protein stability, kinetics, and structure. A recent review summarizes the available computational methods for protein folding.

Energy Landscape of Protein Folding

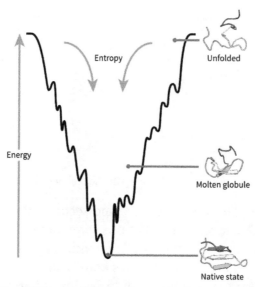

The energy funnel by which an unfolded polypeptide chain assumes its native structure

The protein folding phenomenon was largely an experimental endeavor until the formulation of an energy landscape theory of proteins by Joseph Bryngelson and Peter Wolynes in the late 1980s and early 1990s. This approach introduced the *principle of minimal frustration*. This principle says that nature has chosen amino acid sequences so that the folded state of the protein is very stable. In addition, the undesired interactions between amino acids along the folding pathway are reduced, making the acquisition of the folded state a very fast process. Even though nature has reduced the level of *frustration* in proteins, some degree of it remains up to now as can be observed in the presence of local minima in the energy landscape of proteins. A consequence of these evolutionarily selected sequences is that proteins are generally thought to have globally "funneled energy landscapes" (coined by José Onuchic) that are largely directed toward the native state. This "folding funnel" landscape allows the protein to fold to the native state through any of a large number of pathways and intermediates, rather than being restricted to a single mechanism. The theory is supported by both computational simulations of model proteins and experimental studies, and it has been used to improve methods for protein structure prediction and design. The description of protein folding by the leveling free-energy landscape is also consistent with the 2nd law of thermodynamics. Physically, thinking of landscapes in terms of visualizable potential or total energy surfaces simply with maxima, saddle points, minima, and funnels, rather like geographic landscapes, is perhaps a little misleading. The relevant description is really a high-dimensional phase space in which manifolds might take a variety of more complicated topological forms.

The unfolded polypeptide chain begins at the top of the funnel where it may assume the largest number of unfolded variations and is in its highest energy state. Energy landscapes such as these indicate that there are a large number of initial possibilities, but only a single native state is possible; however, it does not reveal the numerous folding pathways that are possible. A different molecule of the same exact protein may be able to follow marginally different folding pathways, seeking different lower energy intermediates, as long as the same native structure is reached. Different pathways may have different frequencies of utilization depending on the thermodynamic favorability of each pathway. This means that if one pathway is found to be more thermodynamically favorable than another, it is likely to be used more frequently in the pursuit of the native structure. As the protein begins to fold and assume its various conformations, it always seeks a more thermodynamically favorable structure than before and thus continues through the energy funnel. Formation of secondary structures is a strong indication of increased stability within the protein, and only one combination of secondary structures assumed by the polypeptide backbone will have the lowest energy and therefore be present in the native state of the protein. Among the first structures to form once the polypeptide begins to fold are alpha helices and beta turns, where alpha helices can form in as little as 100 nanoseconds and beta turns in 1 microsecond.

There exists a saddle point in the energy funnel landscape where the transition state for a particular protein is found. The transition state in the energy funnel diagram is the conformation that must be assumed by every molecule of that protein if the protein wishes to finally assume the native structure. No protein may assume the native structure without first passing through the transition state. The transition state can be referred to as a variant or premature form of the native state rather than just another intermediary step. The folding of the transition state is shown to be rate-determining, and even though it exists in a higher energy state than the native fold, it greatly resembles the native structure. Within the transition state, there exists a nucleus around which the protein is able to fold, formed by a process referred to as "nucleation condensation" where the structure begins to collapse onto the nucleus.

Two models of protein folding are currently being confirmed:

- The diffusion collision model, in which first a nucleus forms, then the secondary structure, and finally these secondary structures collide and pack tightly together.

- The nucleation-condensation model, in which the secondary and tertiary structures of the protein are made at the same time.

Recent studies have shown that some proteins show characteristics of both of these folding models.

For the most part, scientists have been able to study many identical molecules folding together *en masse*. At the coarsest level, it appears that in transitioning to the native state, a given amino acid sequence takes roughly the same route and proceeds through roughly the same intermediates and transition states. Often folding involves first the establishment of regular secondary and supersecondary structures, in particular alpha helices and beta sheets, and afterward tertiary structure.

Modeling of Protein Folding

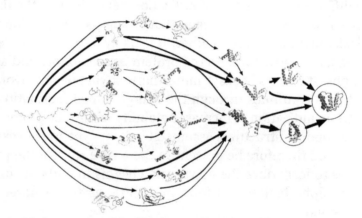

Folding@home uses Markov state models, like the one diagrammed here, to model the possible shapes and folding pathways a protein can take as it condenses from its initial randomly coiled state (left) into its native 3D structure (right)

De novo or *ab initio* techniques for computational protein structure prediction are related to, but strictly distinct from, experimental studies of protein folding. Molecular Dynamics (MD) is an important tool for studying protein folding and dynamics in silico. First equilibrium folding simulations were done using implicit solvent model and umbrella sampling. Because of computational cost, ab initio MD folding simulations with explicit water are limited to peptides and very small proteins. MD simulations of larger proteins remain restricted to dynamics of the experimental structure or its high-temperature unfolding. Long-time folding processes (beyond about 1 millisecond), like folding of small-size proteins (about 50 residues) or larger, can be accessed using coarse-grained models.

The 100-petaFLOP distributed computing project Folding@home created by Vijay Pande's group at Stanford University simulates protein folding using the idle processing time of CPUs and GPUs of personal computers from volunteers. The project aims to understand protein misfolding and accelerate drug design for disease research.

Long continuous-trajectory simulations have been performed on Anton, a massively parallel supercomputer designed and built around custom ASICs and interconnects by D. E. Shaw Research. The longest published result of a simulation performed using Anton is a 2.936 millisecond simulation of NTL9 at 355 K.

Experimental Techniques for Studying Protein Folding

While inferences about protein folding can be made through mutation studies, typically, experimental techniques for studying protein folding rely on the gradual unfolding or folding of proteins and observing conformational changes using standard non-crystallographic techniques.

X-ray Crystallography

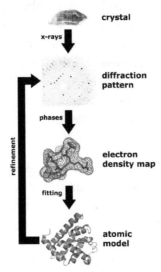

Steps of x-ray crystallography

X-ray crystallography is one of the more efficient and important methods for attempting to decipher the three dimensional configuration of a folded protein. To be able to conduct X-ray crystallography, the protein under investigation must be located inside a crystal lattice. To place a protein inside a crystal lattice, one must have a suitable solvent for crystallization, obtain a pure protein at supersaturated levels in solution, and precipitate the crystals in solution. Once a protein is crystallized, x-ray beams can be concentrated through the crystal lattice which would diffract the beams or shoot them outwards in various directions. These exiting beams are correlated to the specific three-dimensional configuration of the protein enclosed within. The x-rays specifically interact with the electron clouds surrounding the individual atoms within the protein crystal lattice and produce a discernible diffraction pattern. Only by relating the electron density clouds with the amplitude of the x-rays can this pattern be read and lead to assumptions of the phases or phase angles involved that complicate this method. Without the relation established through a mathematical basis known as Fourier transform, the "phase problem" would render predicting the diffraction patterns very difficult. Emerging methods like multiple isomorphous replacement use the presence of a heavy metal ion to diffract the x-rays into a more predictable manner, reducing the number of variables involved and resolving the phase problem.

Fluorescence Spectroscopy

Fluorescence spectroscopy is a highly sensitive method for studying the folding state of proteins. Three amino acids, phenylalanine (Phe), tyrosine (Tyr) and tryptophan (Trp), have intrinsic fluorescence properties, but only Tyr and Trp are used experimentally because their quantum yields are high enough to give good fluorescence signals. Both Trp and Tyr are excited by a wavelength of 280 nm, whereas only Trp is excited by a wavelength of 295 nm. Because of their aromatic character, Trp and Tyr residues are often found fully or partially buried in the hydrophobic core of proteins, at the interface between two protein domains, or at the interface between subunits of oligomeric proteins. In this apolar environment, they have high quantum yields and therefore high fluorescence intensities. Upon disruption of the protein's tertiary or quaternary structure, these side chains become more exposed to the hydrophilic environment of the solvent, and their quantum yields decrease, leading to low fluorescence intensities. For Trp residues, the wavelength of their maximal fluorescence emission also depend on their environment.

Fluorescence spectroscopy can be used to characterize the equilibrium unfolding of proteins by measuring the variation in the intensity of fluorescence emission or in the wavelength of maximal emission as functions of a denaturant value. The denaturant can be a chemical molecule (urea, guanidinium hydrochloride), temperature, pH, pressure, etc. The equilibrium between the different but discrete protein states, i.e. native state, intermediate states, unfolded state, depends on the denaturant value; therefore, the global fluorescence signal of their equilibrium mixture also depends on this value. One thus obtains a profile relating the global protein signal to the denaturant value. The profile of equilibrium unfolding may enable one to detect and identify intermediates of unfolding. General equations have been developed by Hugues Bedouelle to obtain the thermodynamic parameters that characterize the unfolding equilibria for homomeric or heteromeric proteins, up to trimers and potentially tetramers, from such profiles. Fluorescence spectroscopy can be combined with fast-mixing devices such as stopped flow, to measure protein folding kinetics, generate a chevron plot and derive a Phi value analysis.

Circular Dichroism

Circular dichroism is one of the most general and basic tools to study protein folding. Circular dichroism spectroscopy measures the absorption of circularly polarized light. In proteins, structures such as alpha helices and beta sheets are chiral, and thus absorb such light. The absorption of this light acts as a marker of the degree of foldedness of the protein ensemble. This technique has been used to measure equilibrium unfolding of the protein by measuring the change in this absorption as a function of denaturant concentration or temperature. A denaturant melt measures the free energy of unfolding as well as the protein's m value, or denaturant dependence. A temperature melt measures the melting temperature (T_m) of the protein. As for fluorescence spectroscopy, circular-dichroism spectroscopy can be combined with fast-mixing devices such as stopped flow to measure protein folding kinetics and to generate chevron plots.

Vibrational Circular Dichroism of Proteins

The more recent developments of vibrational circular dichroism (VCD) techniques for proteins, currently involving Fourier transform (FFT) instruments, provide powerful means for determining protein conformations in solution even for very large protein molecules. Such VCD studies of proteins are often combined with X-ray diffraction of protein crystals, FT-IR data for protein solutions in heavy water (D_2O), or *ab initio* quantum computations to provide unambiguous structural assignments that are unobtainable from CD.

Protein Nuclear Magnetic Resonance Spectroscopy

Protein folding is routinely studied using NMR spectroscopy, for example by monitoring hydrogen-deuterium exchange of backbone amide protons of proteins in their native state, which provides both the residue-specific stability and overall stability of proteins.

Dual Polarisation Interferometry

Dual polarisation interferometry is a surface-based technique for measuring the optical properties of molecular layers. When used to characterize protein folding, it measures the conformation by determining the overall size of a monolayer of the protein and its density in real time at sub-Angstrom resolution, although real-time measurement of the kinetics of protein folding are limited to processes that occur slower than ~10 Hz. Similar to circular dichroism, the stimulus for folding can be a denaturant or temperature.

Studies of Folding with High Time Resolution

The study of protein folding has been greatly advanced in recent years by the development of fast, time-resolved techniques. Experimenters rapidly trigger the folding of a sample of unfolded protein and observe the resulting dynamics. Fast techniques in use include neutron scattering, ultrafast mixing of solutions, photochemical methods, and laser temperature jump spectroscopy. Among the many scientists who have contributed to the development of these techniques are Jeremy Cook, Heinrich Roder, Harry Gray, Martin Gruebele, Brian Dyer, William Eaton, Sheena Radford, Chris Dobson, Alan Fersht, Bengt Nölting and Lars Konermann.

Proteolysis

Proteolysis is routinely used to probe the fraction unfolded under a wide range of solution conditions (e.g. Fast parallel proteolysis (FASTpp).

Optical Tweezers

Single molecule techniques such as optical tweezers and AFM have been used to understand protein folding mechanisms of isolated proteins as well as proteins with chaperones. Optical tweezers have been used to stretch single protein molecules from their C- and N-termini and unfold them to allow study of the subsequent refolding. The technique allows one to measure folding rates at single-molecule level; for example, optical tweezers have been recently applied to study folding and unfolding of proteins involved in blood coagulation. von Willebrand factor (vWF) is a protein with an essential role in blood clot formation process. It discovered – using single molecule optical tweezers measurement – that calcium-bound vWF acts as a shear force sensor in the blood. Shear force leads to unfolding of the A2 domain of vWF, whose refolding rate is dramatically enhanced in the presence of calcium. Recently, it was also shown that the simple src SH3 domain accesses multiple unfolding pathways under force.

Incorrect Protein Folding and Neurodegenerative Disease

A protein is recognized as misfolded if it cannot achieve the native state of a protein. This can be due to unwanted mutations in the amino acid sequence or could be caused through errors in the folding process. The misfolded protein typically contains beta-sheets that are organized in a polymeric arrangement known as a cross-beta structure. The abnormal proteins that are rich in beta sheet structure become partially resistant to proteolysis. Part of the reason is because beta sheets are stabilized by intermolecular interactions, therefore the misfolded proteins will have a high tendency to form oligomers and larger polymers. Also the accumulation of misfolded proteins can lead to an accumulation of protein aggregates or oligomers in the cell. The increased levels of aggregated proteins in the cell leads to formation of amyloid- like structures which can cause degenerative disorders and cell death. The amyloids are fibrillary structure that contain intermolecular hydrogen bonds, which are highly insoluble, and made from converted protein aggregates. Therefore, the proteasome pathway may not be efficient enough to degrade the misfolded proteins prior to aggregation. Misfolded proteins can interact with one another and form structured aggregates and gain toxicity through intermolecular interactions.

Aggregated proteins are associated with prion-related illnesses such as Creutzfeldt–Jakob disease, bovine spongiform encephalopathy (mad cow disease), amyloid-related illnesses such as Alzheimer's disease and familial amyloid cardiomyopathy or polyneuropathy, as well as intracytoplasmic aggregation diseases such as Huntington's and Parkinson's disease. These age onset degenerative diseases are associated with the aggregation of misfolded proteins into insoluble, extracellular aggregates and/or intracellular inclusions including cross-beta sheet amyloid fibrils. It is not completely clear whether the aggregates are the cause or merely a reflection of the loss of protein homeostasis, the balance between synthesis, folding, aggregation and protein turnover. Recently the European Medicines Agency approved the use of Tafamidis or Vyndaqel (a kinetic stabilizer of tetrameric transthyretin) for the treatment of transthyretin amyloid diseases. This suggests that the process of amyloid fibril formation (and not the fibrils themselves) causes the degeneration of post-mitotic tissue in human amyloid diseases. Misfolding and excessive degradation instead of folding and function leads to a number of proteopathy diseases such as antitrypsin-associated emphysema, cystic fibrosis and the lysosomal storage diseases, where loss of function is the origin of the disorder. While protein replacement therapy has historically been used to correct the latter disorders, an emerging approach is to use pharmaceutical chaperones to fold mutated proteins to render them functional.

Levinthal's Paradox and Kinetics

In 1969, Cyrus Levinthal noted that, because of the very large number of degrees of freedom in an unfolded polypeptide chain, the molecule has an astronomical number of possible conformations. An estimate of 3^{300} or 10^{143} was made in one of his papers. Levinthal's paradox is a thought experiment based on the observation that if a protein were folded by sequentially sampling of all possible conformations, it would take an astronomical amount of time to do so, even if the conformations were sampled at a rapid rate (on the nanosecond or picosecond scale). Based upon the observation that proteins fold much faster than this, Levinthal then proposed that a random conformational search does not occur, and the protein must, therefore, fold through a series of meta-stable intermediate states.

The duration of the folding process varies dramatically depending on the protein of interest. When studied outside the cell, the slowest folding proteins require many minutes or hours to fold primarily due to proline isomerization, and must pass through a number of intermediate states, like checkpoints, before the process is complete. On the other hand, very small single-domain proteins with lengths of up to a hundred amino acids typically fold in a single step. Time scales of milliseconds are the norm and the very fastest known protein folding reactions are complete within a few microseconds.

Protein Separation

In order to analyse different components of a mixture, their separation is necessary. A successful separation needs different separation techniques. There are several physical and chemical methods to separate the different components of a simple mixture but for a complex mixture as, mixture of amino acids or proteins, we need more advanced techniques. Chromatography is the result of advancement of separation and purification techniques. The word Chromatography comes from Greek words chroma chroma "color" and graphein "to write". Literally meaning of Chromatography is "color writing" as initially color of separated component was used for identification. This separation technique credited to a Russian botanist Dr. Michel Tswett (1903). He separated the green plant pigment and showed different components of chlorophyll in form of different colour bands on a $CaCO_3$ column through adsorption. Likewise Richard Kuhn resolves α and β isomers of carotene and proved the importance of adsorption in analytical field (The Nobel Prize in Chemistry 1938).

Column Chromatography

This is the most commonly used mode of chromatography. It can be defined as a separation process involving the uniform percolation of a liquid through a column packed with finely divided material. The stationary phase attached to a matrix (inert insoluble support) is packed in a glass or metal column and mobile phase is passed through it either by gravitational flow or with help of a pump. The selected stationary phase retards the movement of certain components of mobile phase which leads to effective separation of different components. This retardation in movement may be either due to direct interaction of solute component with stationary phase or indirect adsorption of solute component on the surface of stationary phase.

A typical column chromatographic system uses liquid mobile phase, consist of a column, a mobile phase reservoir and delivery system (pump), a detector for identification of separated analytes as they emerge in the effluent from the column, a recorder to maintain the record of analytes in combination with detector and a fraction collector to collect each analyte separately.

The pictorial representation of detector response as a function of elution time or volume is known as chromatogram which consists of a series of peaks, representing the individual analytes. After application on the column the time require to elute by an analyte is known as retention time (t_R) of that analyte. The retention time of any analyte has two components. The first one is time taken by that analyte to cross the column through free spaces between matrixes. This volume is referred as void volume (V_0) and the time taken is called dead time (t_M). The value of dead time (t_M) will be same for all analytes and can be measured by the analyte which does not show any interaction with stationary phase. The second component is the time the analyte retained by the stationary phase and known as adjusted retention time (t_R'). This is the characteristic of analyte and can be given by;

$$t_R' = t_R - t_M$$

The additional time taken by an analyte to elute from the column relative to an excluded analyte that does not interact with the stationary phase is known as capacity factor (k').

Thus, Capacity factor $(k') = t_R - t_M / t_M = t_R'/t_M$ (capacity factor has no units)

Column efficiency and resolution: The efficiency of a chromatographic column is a measure of the capacity of the column to restrain peak dispersion and thus, provide high resolution. The higher the efficiency, the more the peak dispersion is restrained, and the better the column. The column efficiency will vary with the retention of the peak. In capillary columns, the efficiency generally falls as the retention increases and for a packed column the efficiency generally increases with retention.

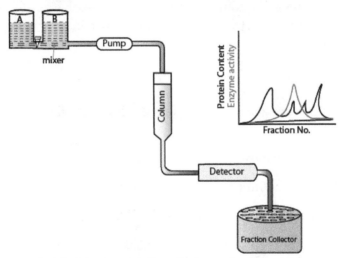

A pictorial representation of column chromatography

The expression for calculating the column efficiency (N) can be derived from the plate theory. Column efficiency is measured in theoretical plates (from the plate theory) and is taken as 16 times the square of the ratio of the retention time (the time between the injection point and the peak maximum) to the peak width at the points of inflection.

$$N = 16 \, t_{R^2} / w^2$$

The height equivalent to the theoretical plate (HETP) or the variance per unit Length of a column is calculated as the ratio of the column length to the column efficiency (number of plates). If the length of the column is L, then the HETP (height equivalent to the theoretical plate) is;

$$HETP = L / N$$

Note: The plate model supposes that the chromatographic column contains a large number of separate layers, called theoretical plates. Separate equilibrations of the sample between the stationary and mobile phase occur in these "plates". The analyte moves down the column by transfer of equilibrated mobile phase from one plate to the next. It is important to remember that the plates do not really exist; they are a figment of the imagination that helps us understand the processes at work in the column. They also serve as a way of measuring column efficiency, either by stating the number of theoretical plates in a column, N (the more plates the better), or by stating the plate height; (the height equivalent to a theoretical plate) (Smaller the HEPT, better the column efficiency).

The number of theoretical plates that a real column possesses can be found by examining a chromatographic peak after elution;

$$N = 5.55\, t_{R^2}/w_{1/2^2} \text{ (where } w_{1/2} \text{ is the peak width at half - height)}$$

Resolution of chromatographic column is the ability to resolve one analyte peak to other. Resolution can be defined as the ratio of the difference in retention time between the two peaks to the mean of their base widths (w_{av}).

$$R_s = 2(t_{RA} - t_{RB})/w_A + w_B = dt_R/w_{av}$$

On the basis of the pressure generated inside the column the liquid column chromatography can be further subdivided as -

- Low pressure liquid chromatography (LPLC) - < 5 Bar.
- Medium pressure liquid chromatography (MPLC) – 6 to 50 Bar.
- High pressure liquid chromatography (HPLC) - > 50 Bar.

In low pressure chromatography, stationary phase are mainly polysaccharides which are mechanically weak. Therefore, even if such particles are produced with small diameters, they would not be sufficiently strong to withstand the high pressure required for high resolution chromatography. Thus, in HPLC silica based particles (5-10 µm) are used. Smaller size of particle (large surface area) reduces size of theoretical chromatographic plat (thus number of plate in a given length is more). Higher number of chromatographic plate results in better resolution.

Types of column chromatography: On the basis of type of interaction between the stationary and the mobile phases the column chromatography can be divided into following types.

1. Adsorption chromatography
2. Partition chromatography
3. Ion-Exchange chromatography
4. Molecular Exclusion (Gel Filtration) chromatography
5. Affinity chromatography

Adsorption Chromatography

Adsorption is a surface phenomenon in which molecules get attached to column particle. The molecule is called adsorbate while column particle is adsorbent. Molecules can be separated based on their adsorption properties. Many inert materials can be used as adsorbents. e.g. alumina, charcoal, calcium carbonate.

Partition Chromatography

Chromatography is a non destructive separation technique for successful separation and purification of individual components of a complex mixture which can not be separated by usual methods due to their similar physical and chemical properties. The basic principal of any chromatography is, how a compound get distributed in between two immiscible phases, a stationary phase and a mobile phase. Those components have higher affinity for the stationary phase are retained longer in the system than those that are distributed selectively in the mobile phase. As a consequence, solutes are eluted from the system as local concentrations in the mobile phase in the order of their increasing distribution coefficients with respect to the stationary phase which leads the separation of different components of mixture.

An analyte remains in equilibrium between the two phases;

$$A_{mobile} \rightleftarrows A_{stationary}$$

The equilibrium constant, K_d is termed the partition coefficient or distribution coefficient, defined as the molar concentration of analyte in the stationary phase divided by the molar concentration of the analyte in the mobile phase. It explains about the distribution of a compound in between two phases. Suppose there are two phases A & B (immiscible to each other) and these are present together. When a compound X is mixed, it distributes itself in these two phases and the concentration of compound in phase A is X_A and in phase B is X_B. The distribution coefficient of X can be expressed as,

$$K_d = X_A / X_B \quad (K_d \text{ is a constant at a particular temperature})$$

Factors Affecting the Magnitude of the Distribution Coefficient (K_d)

The magnitude of (K_d) is determined by the relative affinity of the solute for the two phases. Those solutes interacting more strongly with the stationary phase will exhibit a high distribution coefficient and will be retained longer in the chromatographic system. Molecular interaction results from intermolecular forces of which there are three basic types. (a) Dispersion forces (b) Polar forces (c) Ionic forces.

If the compound X has same affinity for both the phase, it gets distributed equally in both the phases but compound Y has differential affinity for two phases and accordingly gets distributed in stationary and mobile phase.

Protein Chemistry

The central dogma of life places translation process and protein products at the final stage. Protein play most important role in determining the role of the cell in the system. The DNA sequence determines the sequence of amino acids in the protein, but it is actually the modifications in the proteins resulting from alternate splicing and post posttranslational modification, which finally dictates the physiological function of the protein. The structure and the function of the proteins are closely associated. In fact, it is the 3D structure of the protein, which governs the function of the protein. Hence an approach should be available so that all information regarding proteins can be obtained in high throughput manner.

Proteomics differs from the conventional protein chemistry approach of identification, structural and functional elucidation of proteins, in many aspects. Proteomics aims to decipher the protein properties from a global perspective, not taking one protein at a time. This approach becomes a holistic model, while studying the system, because in a system the protein in question is not alone. It is associated and interacting with several other proteins and bio-molecules, which together determine its role in the system. These two approaches, protein chemistry and proteomics, also differ in the techniques that are employed for these approaches as well as scale of usability. While traditional protein chemistry also employed techniques like electrophoresis and mass spectrometry, it is the advancement of these techniques, which led to the divergence from protein chemistry to proteomics. The sensitivity, resolvability, robustness and the high throughput approaches made the transition possible. The completion of genome sequence projects and advancement in bioinformatics and microarrays were also major catalysts for the advancement of proteomics. These advancements in area of protein chemistry and biology gave rise to this new discipline, proteomics, and with continuous evolution of several advanced technologies, this field is continuously advancing.

Evolution of Proteomics

The number of protein coding genes in the human genome is approximately 5% of the entire genome. It is an astonishing fact that how these 5% of the genes account for the entire diversity of proteins in the cell. It is the dynamic properties of proteins, which lead to several diseases. For example, cell cycle regulating proteins or cyclin dependent kinases phosphorylate activating the cyclins, which promote the cell cycle by allowing DNA replication to pass through the check points. Proteomics emerged from protein chemistry solely because of the advancement in the existing technologies.

Advancement in Mass Spectrometry

Mass spectrometry has been into existence during last several decades, since the days of protein chemistry. Mass spectrometry ideally measures the mass of an analyte by producing charged molecular species in vacuum, and their separation by magnetic and electric fields based on mass to charge (m/z) ratio. The pre-requisite for a mass spectrometry analysis is the generation of ionized particles. However, proteins being large soluble polymers of amino acids could not be ionized by the conventional gas chromatography without fragmenting it into constituent amino acids. This for the time being limited the usage of mass spectrometry in protein chemistry. However, with major discoveries of soft ionization techniques like MALDI and ESI, the mass spectrometry became a robust analytical technique for protein study.

With advancement in ionization techniques, sophisticated mass analyzers and detectors, the mass analysis of as low as 1 ppm with excellent resolving power is possible. The sensitivity of the mass spectrometry also increased, making it possible to detect the attomole of substances. The tandem mass spectrometry involving two mass analyzers like Tof-Tof/ Q-Tof etc. helped in protein sequencing. The Edman degradation process of amino acid sequencing though was extremely useful, was not high throughput as it could sequence a stretch of maximum 40 amino acids, whereas mass spectrometry could do it for large proteins. The ability of mass spectrometry for relative and absolute quantitation of proteins by employing iTRAQ, ICAT and SILAC labels have definitely advanced the field of quantitative proteomics.

Advancement in Electrophoresis

Electrophoresis refers to the process of separation of charged particles under the influence of an external electric field. Electrophoretic separation of proteins had been used widely for quite a long time; however, recent advances in electrophoretic techniques in the form of protein separation, staining and detection have advanced this technique for further usage in proteomics.

The first advancement came in the transition from tube gels to immobilized pH gradient strips. Earlier, isoelectric focusing was performed using tube gels, which had biggest disadvantage of their stability. The pH would often change and result in erroneous results. The tube gels suffered from extremes of variations, discontinuity in the entire gel, which led to its breakage when concentrated samples were added. Prof. Angelika Gorg has made a remarkable contribution in the field of electrophoresis by substituting the tube gels with immobilized pH gradient strips. These were Acrylamide coated plastic strips containing immobilins of various pH spread across them. The biggest advantage of these strips was in the stability of the pH ampholytes inside the gel. As a result gel to gel variations got reduced and the physical stability of the strips were enhanced, as much as, they can be stored after the first dimension for a week at -200C before the second dimension can be done.

Staining techniques also increased the usability of gel electrophoresis in proteomic studies. The conventional coomassie brilliant blue dye had the limitation of 40 mg protein requirement. Although silver staining increased the sensitivity but it had its own limitation in terms of background staining and incompatibility with mass spectrometry for the protein identification. To overcome these problems, cyanine dyes were introduced for staining which have extreme sensitivity as well as the specificity. The cyanine dyes interact with the protein with their hydroxysuccinamide residues and fluoresce when subjected to light of appropriate wavelength. This property of cyanine dyes is utilized into developing a technique known as DIGE (Difference in gel electrophoresis). The biggest advantage of DIGE was in its ability to resolve gel-to-gel variation. Since an internal standard is included in the gel, there is no need to perform electrophoresis on many samples. The sensitivity of the dyes, coupled with the latest software could thus even check the smallest amount of differential expression level of a protein with reliability. Although electrophoretic techniques such as 2DE is now considered to be primitive in proteomics; however, quantitative approaches such as 2D-DIGE still hold good place for quantitative proteomic analysis.

Protein Mass Spectrometry

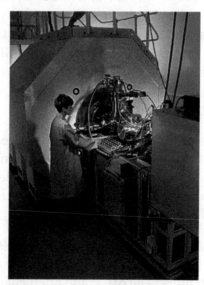

A mass spectrometer used for high throughput protein analysis

Protein mass spectrometry refers to the application of mass spectrometry to the study of proteins. Mass spectrometry is an important method for the accurate mass determination and characterization of proteins, and a variety of methods and instrumentations have been developed for its many uses. Its applications include the identification of proteins and their post-translational modifications, the elucidation of protein complexes, their subunits and functional interactions, as well as the global measurement of proteins in proteomics. It can also be used to localize proteins to the various organelles, and determine the interactions between different proteins as well as with membrane lipids.

The two primary methods used for the ionization of protein in mass spectrometry are electrospray ionization (ESI) and matrix-assisted laser desorption/ionization (MALDI). These ionization techniques are used in conjunction with mass analyzers such as tandem mass spectrometry. In general, the protein are analyzed either in a "top-down" approach in which proteins are analyzed intact, or a "bottom-up" approach in which protein are first digested into fragments. An intermediate "middle-down" approach in which larger peptide fragments are analyzed may also sometimes be used.

History

The application of mass spectrometry to study proteins became popularized in the 1980s after the development of MALDI and ESI. These ionization techniques have played a significant role in the characterization of proteins. (MALDI) Matrix-assisted laser desorption ionization was coined in the late 80's by Franz Hillenkamp and Michael Karas. Hillenkamp, Karas and their fellow researchers were able to ionize the amino acid alanine by mixing it with the amino acid tryptophan and irradiated with a pulse 266 nm laser. Though important, the breakthrough did not come until 1987. In 1987, Koichi Tanaka used the "ultra fine metal plus liquid matrix method" and ionized biomolecules the size of 34,472 Da protein carboxypeptidase-A.

In 1968, Malcolm Dole reported the first use of electrospray ionization with mass spectrometry. Around the same time MALDI became popularized, John Bennett Fenn was cited for the development of electrospray ionization. Koichi Tanaka received the 2002 Nobel Prize in Chemistry alongside John Fenn, and Kurt Wüthrich "for the development of methods for identification and structure analyses of biological macromolecules." These ionization methods have greatly facilitated the study of proteins by mass spectrometry. Consequently, protein mass spectrometry now plays a leading role in protein characterization.

Methods and Approaches

Techniques

Mass spectrometry of proteins requires that the proteins in solution or solid state be turned into an ionized form in the gas phase before they are injected and accelerated in an electric or magnetic field for analysis. The two primary methods for ionization of proteins are electrospray ionization (ESI) and matrix-assisted laser desorption/ionization (MALDI). In electrospray, the ions are created from proteins in solution, and it allows fragile molecules to be ionized intact, sometimes preserving non-covalent interactions. In MALDI, the proteins are embedded within a matrix normally in a solid form, and ions are created by pulses of laser light. Electrospray produces more multiply-charged ions than MALDI, allowing for measurement of high mass protein and better fragmentation for identification, while MALDI is fast and less likely to be affected by contaminants, buffers and additives.

Whole-protein mass analysis is primarily conducted using either time-of-flight (TOF) MS, or Fourier transform ion cyclotron resonance (FT-ICR). These two types of instrument are preferable here because of their wide mass range, and in the case of FT-ICR, its high mass accuracy. Electrospray ionization of a protein often results in generation of multiple charged species of $800 < m/z < 2000$ and the resultant spectrum can be deconvoluted to determine the protein's average mass to within 50 ppm or better using TOF or ion-trap instruments.

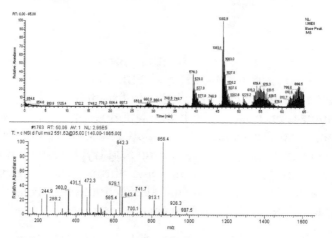

Chromatography trace and MS/MS spectra of a peptide

Mass analysis of proteolytic peptides is a popular method of protein characterization, as cheaper instrument designs can be used for characterization. Additionally, sample preparation is easier once whole proteins have been digested into smaller peptide fragments. The most widely used instrument for peptide mass analysis are the MALDI-TOF instruments as they permit the acquisition of peptide mass fingerprints (PMFs) at high pace (1 PMF can be analyzed in approx. 10 sec). Multiple stage quadrupole-time-of-flight and the quadrupole ion trap also find use in this application.

Tandem mass spectrometry (MS/MS) is used to measure fragmentation spectra and identify proteins at high speed and accuracy. Collision-induced dissociation is used in mainstream applications to generate a set of fragments from a specific peptide ion. The fragmentation process primarily gives rise to cleavage products that break along peptide bonds. Because of this simplicity in fragmentation, it is possible to use the observed fragment masses to match with a database of predicted masses for one of many given peptide sequences. Tandem MS of whole protein ions has been investigated recently using electron capture dissociation and has demonstrated extensive sequence information in principle but is not in common practice.

Approaches

In keeping with the performance and mass range of available mass spectrometers, two approaches are used for characterizing proteins. In the first, intact proteins are ionized by either of the two techniques described above, and then introduced to a mass analyzer. This approach is referred to as "top-down" strategy of protein analysis as it involves starting with the whole mass and then pulling it apart. The top-down approach however is mostly limited to low-throughput single-protein studies due to issues involved in handling whole proteins, their heterogeneity and the complexity of their analyses.

In the second approach, referred to as the "bottom-up" MS, proteins are enzymatically digested into smaller peptides using a protease such as trypsin. Subsequently, these peptides are introduced into the mass spectrometer and identified by peptide mass fingerprinting or tandem mass spectrometry. Hence, this approach uses identification at the peptide level to infer the existence of proteins pieced back together with *de novo* repeat detection. The smaller and more uniform fragments are easier to analyze than intact proteins and can be also determined with high accuracy, this "bottom-up" approach is therefore the preferred method of studies in proteomics. A further approach that is beginning to be useful is the intermediate "middle-down" approach in which proteolytic peptides larger than the typical tryptic peptides are analyzed.

Protein and Peptide Fractionation

Proteins of interest are usually part of a complex mixture of multiple proteins and molecules, which co-exist in the biological medium. This presents two significant problems. First, the two ionization techniques used for large molecules only work well when the mixture contains roughly equal amounts of constituents, while in biological samples, different proteins tend to be present in widely differing amounts. If such a mixture is ionized using electrospray or MALDI, the more abundant species have a tendency to "drown" or suppress signals from less abundant ones. Second, mass spectrum from a complex mixture is very difficult to interpret due to the overwhelming number of mixture components. This is exacerbated by the fact that enzymatic digestion of a protein gives rise to a large number of peptide products.

In light of these problems, the methods of one- and two-dimensional gel electrophoresis and high performance liquid chromatography are widely used for separation of proteins. The first method fractionates whole proteins via two-dimensional gel electrophoresis. The first-dimension of 2D gel is isoelectric focusing (IEF). In this dimension, the protein is separated by its isoelectric point (pI) and the second-dimension is SDS-polyacrylamide gel electrophoresis (SDS-PAGE). This dimension separates the protein according to its molecular weight. Once this step is completed in-gel digestion occurs. In some situations, it may be necessary to combine both of these techniques. Gel spots identified on a 2D Gel are usually attributable to one protein. If the identity of the protein is desired, usually the method of in-gel digestion is applied, where the protein spot of interest is excised, and digested proteolytically. The peptide masses resulting from the digestion can be determined by mass spectrometry using peptide mass fingerprinting. If this information does not allow unequivocal identification of the protein, its peptides can be subject to tandem mass spectrometry for *de novo* sequencing.

Small changes in mass and charge can be detected with 2D-PAGE. The disadvantages with this technique are its small dynamic range compared to other methods, some proteins are still difficult to separate due to their acidity, basicity, hydrophobicity, and size (too large or too small).

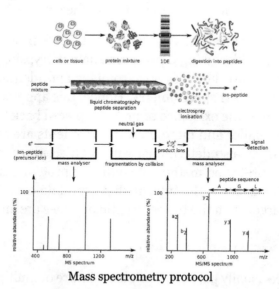

Mass spectrometry protocol

The second method, high performance liquid chromatography is used to fractionate peptides after enzymatic digestion. Characterization of protein mixtures using HPLC/MS is also called shotgun proteomics and MuDPIT (Multi-Dimensional Protein Identification Technology). A peptide mixture that results from digestion of a protein mixture is fractionated by one or two steps of liquid chromatography. The eluent from the chromatography stage can be either directly introduced to the mass spectrometer through electrospray ionization, or laid down on a series of small spots for later mass analysis using MALDI.

Applications

Protein Identification

There are two main ways MS is used to identify proteins. Peptide mass fingerprinting uses the masses of proteolytic peptides as input to a search of a database of predicted masses that would arise from digestion of a list of known proteins. If a protein sequence in the reference list gives rise to a significant number of predicted masses that match the experimental values, there is some evidence that this protein was present in the original sample. Purification steps therefore limit the throughput of the peptide mass fingerprinting approach. Peptide mass fingerprinting can be achieved with MS/MS.

MS is also the preferred method for the identification of post-translational modifications in proteins as it is more advantageous than other approaches such as the antibody-based methods.

De Novo (Peptide) Sequencing

De novo peptide sequencing for mass spectrometry is typically performed without prior knowledge of the amino acid sequence. It is the process of assigning amino acids from peptide fragment masses of a protein. *De novo* sequencing has proven successful for confirming and expanding upon results from database searches.

As *de novo* sequencing is based on mass and some amino acids have identical masses (e.g. leucine and isoleucine), accurate manual sequencing can be difficult. Therefore, it may be necessary to utilize a sequence homology search application to work in tandem between a database search and *de novo* sequencing to address this inherent limitation.

Database searching has the advantage of quickly identifying sequences, provided they have already been documented in a database. Other inherent limitations of database searching include sequence modifications/mutations (some database searches do not adequately account for alterations to the 'documented' sequence, thus can miss valuable information), the unknown (if a sequence is not documented, it will not be found), false positives, and incomplete and corrupted data.

An annotated peptide spectral library can also be used as a reference for protein/peptide identification. It offers the unique strength of reduced search space and increased specificity. The limitations include spectra not included in the library will not be identified, spectra collected from different types of mass spectrometers can have quite distinct features, and reference spectra in the library may contain noise peaks, which may lead to false positive identifications. A number of different algorithmic approaches have been described to identify peptides and proteins from tandem mass spectrometry (MS/MS), peptide *de novo* sequencing and sequence tag-based searching.

Protein Quantitation

Several recent methods allow for the quantitation of proteins by mass spectrometry (quantitative proteomics). Typically, stable (e.g. non-radioactive) heavier isotopes of carbon (^{13}C) or nitrogen (^{15}N) are incorporated into one sample while the other one is labeled with corresponding light isotopes (e.g. ^{12}C and ^{14}N). The two samples are mixed before the analysis. Peptides derived from the different samples can be distinguished due to their mass difference. The ratio of their peak intensities corresponds to the relative abundance ratio of the peptides (and proteins). The most popular methods for isotope labeling are SILAC (stable isotope labeling by amino acids in cell culture), trypsin-catalyzed ^{18}O labeling, ICAT (isotope coded affinity tagging), iTRAQ (isobaric tags for relative and absolute quantitation). "Semi-quantitative" mass spectrometry can be performed without labeling of samples. Typically, this is done with MALDI analysis (in linear mode). The peak intensity, or the peak area, from individual molecules (typically proteins) is here correlated to the amount of protein in the sample. However, the individual signal depends on the primary structure of the protein, on the complexity of the sample, and on the settings of the instrument. Other types of "label-free" quantitative mass spectrometry, uses the spectral counts (or peptide counts) of digested proteins as a means for determining relative protein amounts.

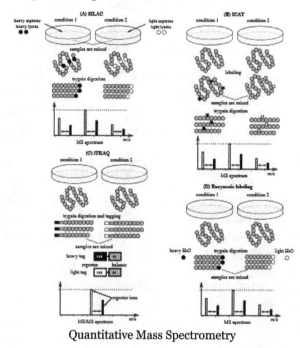

Quantitative Mass Spectrometry

Protein Structure Determination

Characteristics indicative of the 3-dimensional structure of proteins can be probed with mass spectrometry in various ways. By using chemical crosslinking to couple parts of the protein that are close in space, but far apart in sequence, information about the overall structure can be inferred. By following the exchange of amide protons with deuterium from the solvent, it is possible to probe the solvent accessibility of various parts of the protein. Hydrogen-deuterium exchange mass spectrometry has been used to study proteins and their conformations for over 20 years. This type of protein structural analysis can be suitable for proteins that are challenging for other structural methods. Another interesting avenue in protein structural studies is laser-induced covalent labeling. In this technique, solvent-exposed sites of the protein are modified by hydroxyl radicals. Its combination with rapid mixing has been used in protein folding studies.

Biomarkers

The FDA defines a biomarker as, "A characteristic that is objectively measured and evaluated as an indicator of normal biologic processes, pathogenic processes, or pharmacologic responses to a therapeutic intervention". It is hypothesized that mass spectrometry enables the discovery of candidates for biomarkers.

Proteogenomics

In what is now commonly referred to as proteogenomics, peptides identified with mass spectrometry are used for improving gene annotations (for example, gene start sites) and protein annotations. Parallel analysis of the genome and the proteome facilitates discovery of post-translational modifications and proteolytic events, especially when comparing multiple species.

Quantitative Proteomics

Quantitative proteomics is an analytical chemistry technique for determining the amount of proteins in a sample. Rather than just providing lists of proteins identified in a certain sample, quantitative proteomics yields information about differences between samples. For example, this approach can be used to compare samples from healthy and diseased patients. The methods for protein identification are identical to those used in general (i.e. qualitative) proteomics, but include quantification as an additional dimension. Quantitative proteomics is mainly performed by two-dimensional gel electrophoresis (2-DE) or mass spectrometry (MS). However, a recent developed method of Quantitative Dot Blot (QDB) analysis is able to measure both the absolute and relative quantity of an individual proteins in the sample in high throughput format, thus open a new direction for proteomic research. In contrast to 2-DE, which requires MS for the downstream protein identification, MS technology can identify and quantify the changes.

Technologies

Mass spectrometry (MS) and two-dimensional gel electrophoresis (2-DE) represent the main technologies for quantitative proteomics with advantages and disadvantages. 2-DE provides information about the protein quantity, charge, and mass of the intact protein. It has limitations for the analysis of proteins larger than 150 kDa or smaller than 5kDa and low solubility proteins. Quantitative MS has higher sensitivity but does not provide information about the intact protein.

Classical 2-DE based on post-electrophoretic dye staining has limitations: at least three technical replicates are required to verify the reproducibility. Difference gel electrophoresis (DIGE) uses fluorescence-based labeling of the proteins prior to separation has increased the precision of quantification as well as the sensitivity in the protein detection. Therefore, DIGE represents the current main approach for the 2-DE based study of proteomes.

For quantitative MS, a commonly applied approach is isotope-coded affinity tags (ICAT), which uses two reagents with heavy and light isotopes, respectively, and a biotin affinity tag to modify cysteine containing peptides. This technology has been used to label whole *Saccharomyces cerevisiae* cells, and, in conjunction with mass spectrometry, helped lay the foundation of quantitative proteomics.

Discovery and Targeted Proteomics

Strategies to improve the sensitivity and scope of proteomic analysis often require large sample quantities and multi-dimensional fractionation, which sacrifices throughput. Alternatively, efforts to improve the sensitivity and throughput of protein quantification limit the number of peptides that can be monitored per MS run. For this reason, proteomics research is typically divided into two categories: discovery and targeted proteomics. Discovery proteomics optimizes protein identification by spending more time and effort per sample and reducing the number of samples analyzed. In contrast, targeted proteomics strategies limit the number of features that will be monitored and then optimize the chromatography, instrument tuning and acquisition methods to achieve the highest sensitivity and throughput for hundreds or thousands of samples.

Relative and Absolute Quantification

Mass spectrometry is not inherently quantitative because of differences in the ionization efficiency and/or detectability of the many peptides in a given sample, which has sparked the development of methods to determine relative and absolute abundance of proteins in samples. The intensity of a peak in a mass spectrum is not a good indicator of the amount of the analyte in the sample, although differences in peak intensity of the *same* analyte between multiple samples accurately reflect relative differences in its abundance.

Label-free Quantification

One approach for relative quantification is to separately analyze samples by MS and compare the spectra to determine peptide abundance in one sample relative to another, as in Label-free quantification strategies.

Stable Isotope Labels

An approach for relative quantification that is more costly and time-consuming, though less sensitive to experimental bias than label-free quantification, entails labeling the samples with stable isotope labels that allow the mass spectrometer to distinguish between identical proteins in separate samples. One type of label, isotopic tags, consist of stable isotopes incorporated into protein crosslinkers that causes a known mass shift of the labeled protein or peptide in the mass spectrum. Differentially labeled samples are combined and analyzed together, and the differences in the peak intensities of the isotope pairs accurately reflect difference in the abundance of the corresponding proteins.

Absolute proteomic quantification using isotopic peptides entails spiking known concentrations of synthetic, heavy isotopologues of target peptides into an experimental sample and then performing LC-MS/MS. As with relative quantification using isotopic labels, peptides of equal chemistry co-elute and are analyzed by MS simultaneously.

Unlike relative quantification, though, the abundance of the target peptide in the experimental sample is compared to that of the heavy peptide and back-calculated to the initial concentration of the standard using a pre-determined standard curve to yield the absolute quantification of the target peptide.

Relative quantification methods include isotope-coded affinity tags (ICAT), isobaric labeling (tandem mass tags (TMT) and isobaric tags for relative and absolute quantification (iTRAQ)), label-free quantification Metal-coded tags (MeCAT), N-terminal labelling, stable isotope labeling with amino acids in cell culture (SILAC), and Terminal amine isotopic labeling of substrates (TAILS).

Absolute quantification is performed using selected reaction monitoring (SRM).

MeCAT can be used in combination with element mass spectrometry ICP-MS allowing first-time absolute quantification of the metal bound by MeCAT reagent to a protein or biomolecule. Thus it is possible to determine the absolute amount of protein down to attomol range using external calibration by metal standard solution. It is compatible to protein separation by 2D electrophoresis and chromatography in multiplex experiments. Protein identification and relative quantification can be performed by MALDI-MS/MS and ESI-MS/MS.

Mass spectrometers have a limited capacity to detect low-abundance peptides in samples with a high dynamic range. The limited duty cycle of mass spectrometers also restricts the collision rate, resulting in an undersampling Sample preparation protocols represent sources of experimental bias.

References

- Fruton JS (1990). "Chapter 5- Emil Fischer and Franz Hofmeister". Contrasts in Scientific Style: Research Groups in the Chemical and Biochemical Sciences,. 191. American Philosophical Society. pp. 163–165. ISBN 0-87169-191-4

- Hatem, Salama Mohamed Ali (2006). "Gas chromatographic determination of Amino Acid Enantiomers in tobacco and bottled wines". University of Giessen. Retrieved 17 November 2008

- Wagner I, Musso H (November 1983). "New Naturally Occurring Amino Acids". Angewandte Chemie International Edition in English. 22 (11): 816–28. doi:10.1002/anie.198308161

- McCoy RH, Meyer CE, Rose WC (1935). "Feeding Experiments with Mixtures of Highly Purified Amino Acids. VIII. Isolation and Identification of a New Essential Amino Acid". Journal of Biological Chemistry. 112: 283–302

- Elmore DT, Barrett GC (1998). Amino acids and peptides. Cambridge, UK: Cambridge University Press. pp. 48–60. ISBN 0-521-46827-2

- "Nomenclature and Symbolism for Amino Acids and Peptides". IUPAC-IUB Joint Commission on Biochemical Nomenclature. 1983. Archived from the original on 9 October 2008. Retrieved 17 November 2008

- Xie J, Schultz PG (December 2005). "Adding amino acids to the genetic repertoire". Current Opinion in Chemical Biology. 9 (6): 548–54. PMID 16260173. doi:10.1016/j.cbpa.2005.10.011

- Young VR (August 1994). "Adult amino acid requirements: the case for a major revision in current recommendations". The Journal of Nutrition. 124 (8 Suppl): 1517S–1523S. PMID 8064412
- Jones RC, Buchanan BB, Gruissem W (2000). Biochemistry & molecular biology of plants. Rockville, Md: American Society of Plant Physiologists. pp. 371–2. ISBN 0-943088-39-9
- Cowtan K (2001). "Phase Problem in X-ray Crystallography, and Its Solution" (PDF). Encyclopedia of Life Sciences. Macmillan Publishers Ltd, Nature Publishing Group. Retrieved November 3, 2016
- Gromer S, Urig S, Becker K (January 2004). "The thioredoxin system--from science to clinic". Medicinal Research Reviews. 24 (1): 40–89. PMID 14595672. doi:10.1002/med.10051
- Reeds PJ (July 2000). "Dispensable and indispensable amino acids for humans". The Journal of Nutrition. 130 (7): 1835S–40S. PMID 10867060
- Hausman, Robert E., Cooper, Geoffrey M. (2004). The cell: a molecular approach. Washington, D.C: ASM Press. p. 51. ISBN 0-87893-214-3
- "Structural Biochemistry/Proteins/Protein Folding - Wikibooks, open books for an open world". en.wikibooks.org. Retrieved 2016-11-05
- Phillips SM (May 2014). "A brief review of critical processes in exercise-induced muscular hypertrophy". Sports Medicine. 44 Suppl 1: S71–7. PMC 4008813 . PMID 24791918. doi:10.1007/s40279-014-0152-3
- Stegink LD (July 1987). "The aspartame story: a model for the clinical testing of a food additive". The American Journal of Clinical Nutrition. 46 (1 Suppl): 204–15. PMID 3300262
- Alberts B, Johnson A, Lewis J, Raff M, Roberts K, Walters P (2002). "The Shape and Structure of Proteins". Molecular Biology of the Cell; Fourth Edition. New York and London: Garland Science. ISBN 0-8153-3218-1
- Pratt C, Cornely K (2004). "Thermodynamics". Essential Biochemistry. Wiley. ISBN 978-0-471-39387-0. Retrieved 2016-11-26
- Turner BL, Harborne JB (1967). "Distribution of canavanine in the plant kingdom". Phytochemistry. 6 (6): 863–66. doi:10.1016/S0031-9422(00)86033-1

Electrophoresis: An Integrated Study

Science and technology have undergone rapid developments in the past decade which has resulted in the discovery of significant techniques in the separation of proteins. These have been extensively detailed in this chapter. Electrophoresis is another method for separation of protein, which is more powerful and accurate than chromatography. It works on the principle that charged particles move towards oppositely charged electrode in a fluid. This chapter discusses in detail the theories and methodologies related to electrophoresis.

Electrophoresis

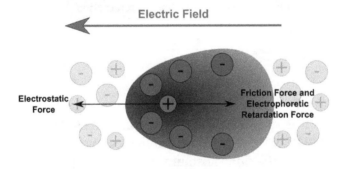

Illustration of electrophoresis

Electrophoresis is the motion of dispersed particles relative to a fluid under the influence of a spatially uniform electric field. This electrokinetic phenomenon was observed for the first time in 1807 by Ferdinand Frederic Reuss (Moscow State University), who noticed that the application of a constant electric field caused clay particles dispersed in water to migrate. It is ultimately caused by the presence of a charged interface between the particle surface and the surrounding fluid. It is the basis for a number of analytical techniques used in chemistry for separating molecules by size, charge, or binding affinity.

Electrophoresis of positively charged particles (cations) is called cataphoresis, while electrophoresis of negatively charged particles (anions) is called anaphoresis. Electrophoresis is a technique used in laboratories in order to separate macromolecules based on size. The technique applies a negative charge so proteins move towards a positive charge. This is used for both DNA and RNA analysis. Polyacrylamide gel electrophoresis (PAGE) has a clearer resolution than agarose and is more suitable for quantitative analysis. In this technique DNA foot-printing can identify how proteins bind to DNA.

It can be used to separate proteins by size, density and purity. It can also be used for plasmid analysis, which develops our understanding of bacteria becoming resistant to antibiotics.

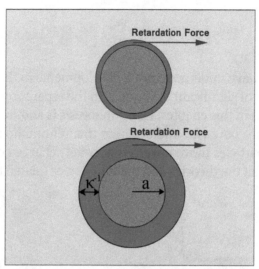

Illustration of electrophoresis retardation

History

Theory

Suspended particles have an electric surface charge, strongly affected by surface adsorbed species, on which an external electric field exerts an electrostatic Coulomb force. According to the double layer theory, all surface charges in fluids are screened by a diffuse layer of ions, which has the same absolute charge but opposite sign with respect to that of the surface charge. The electric field also exerts a force on the ions in the diffuse layer which has direction opposite to that acting on the surface charge. This latter force is not actually applied to the particle, but to the ions in the diffuse layer located at some distance from the particle surface, and part of it is transferred all the way to the particle surface through viscous stress. This part of the force is also called electrophoretic retardation force. When the electric field is applied and the charged particle to be analyzed is at steady movement through the diffuse layer, the total resulting force is zero :

$$F_{tot} = 0 = F_{el} + F_f + F_{ret}$$

Considering the drag on the moving particles due to the viscosity of the dispersant, in the case of low Reynolds number and moderate electric field strength E, the drift velocity of a dispersed particle v is simply proportional to the applied field, which leaves the electrophoretic mobility μ_e defined as:

$$\mu_e = \frac{v}{E}$$

The most well known and widely used theory of electrophoresis was developed in 1903 by Smoluchowski:

$$\mu_e = \frac{\varepsilon_r \varepsilon_0 \zeta}{\eta},$$

where ε_r is the dielectric constant of the dispersion medium, ε_0 is the permittivity of free space ($C^2\ N^{-1}\ m^{-2}$), η is dynamic viscosity of the dispersion medium (Pa s), and ζ is zeta potential (i.e., the electrokinetic potential of the slipping plane in the double layer).

The Smoluchowski theory is very powerful because it works for dispersed particles of any shape at any concentration. It has limitations on its validity. It follows, for instance, because it does not include Debye length κ^{-1}. However, Debye length must be important for electrophoresis, as follows immediately from the Figure. Increasing thickness of the double layer (DL) leads to removing the point of retardation force further from the particle surface. The thicker the DL, the smaller the retardation force must be.

Detailed theoretical analysis proved that the Smoluchowski theory is valid only for sufficiently thin DL, when particle radius a is much greater than the Debye length:

$$a\kappa \gg 1.$$

This model of "thin double layer" offers tremendous simplifications not only for electrophoresis theory but for many other electrokinetic theories. This model is valid for most aqueous systems, where the Debye length is usually only a few nanometers. It only breaks for nano-colloids in solution with ionic strength close to water.

The Smoluchowski theory also neglects the contributions from surface conductivity. This is expressed in modern theory as condition of small Dukhin number:

$$Du \ll 1.$$

In the effort of expanding the range of validity of electrophoretic theories, the opposite asymptotic case was considered, when Debye length is larger than particle radius:

$$a\kappa < 1.$$

Under this condition of a "thick double layer", Hückel predicted the following relation for electrophoretic mobility:

$$\mu_e = \frac{2\varepsilon_r \varepsilon_0 \zeta}{3\eta}.$$

This model can be useful for some nanoparticles and non-polar fluids, where Debye length is much larger than in the usual cases.

There are several analytical theories that incorporate surface conductivity and eliminate the restriction of a small Dukhin number, pioneered by Overbeek. and Booth. Modern, rigorous theories valid for any Zeta potential and often any aκ stem mostly from Dukhin–Semenikhin theory. In the thin double layer limit, these theories confirm the numerical solution to the problem provided by O'Brien and White.

Gel-based Proteomics

Electrophoresis is the widely used technique for protein separation and works on the principle of migration of charged molecules in a gel matrix towards the oppositely charged electrode, under the influence of an applied electric field. It is a powerful technique for finer protein separation and visualization of these separated proteins. Electrophoresis was invented by Prof. Tiselius in 1930 as the moving boundary method to study electrophoresis of proteins. It has been extensively used since then and numerous advancements have been brought about in this technique.

One Dimensional Gel Electrophoresis relies on the principle of separation of protein molecules on the basis of their charge to mass ratio and molecular weight. The low molecular weight proteins are able to migrate larger distances on the gel as compared to the higher molecular weight proteins. Polyacrylamide gels are formed from the polymerization of two compounds, acrylamide and N, N1-methylene- bis-acrylamide (Bis, for short). Bis is a cross-linking agent for the gels. The polymerization is initiated by the addition of ammonium persulfate along with either β-dimethyl amino-propionitrile (DMAP) or N,N,N1,N1,- tetramethylethylenediamine (TEMED). The gels are neutral, hydrophilic and three-dimensional networks of long hydrocarbons cross-linked by methylene groups. The commonly employed 1DE techniques include SDS-PAGE and NATIVE PAGE.

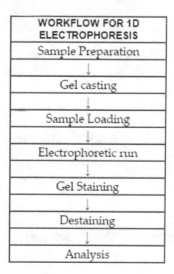

■ Acrylamide: matrix or gelling agent	■ APS: Ammonium Per Sulfate; Initiates polymerization
■ Bis-acrylamide: Cross linking agent	■ TEMED: N,N,N1,N1,- tetramethylethylenediamine; Stabilizes the free radicals and promotes polymerization
■ SDS: Sodium Dodecyl Sulfate, Anionic detergent which imparts uniform negative charge on all the protein molecules which undergo separation	
	■ β-Mercaptoethanol: Acts as a reducing agent and breaks the Disulfide bonds.

Polyacrylamide Gel Electrophoresis

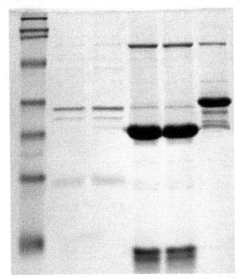

Picture of an SDS-PAGE. The molecular markers (ladder) are in the left lane

Polyacrylamide gel electrophoresis (PAGE), describes a technique widely used in biochemistry, forensics, genetics, molecular biology and biotechnology to separate biological macromolecules, usually proteins or nucleic acids, according to their electrophoretic mobility. Mobility is a function of the length, conformation and charge of the molecule.

As with all forms of gel electrophoresis, molecules may be run in their native state, preserving the molecules' higher-order structure. This method is called Native-PAGE. Alternatively, a chemical denaturant may be added to remove this structure and turn the molecule into an unstructured linear chain whose mobility depends only on its length and mass-to-charge ratio. This procedure is called SDS-PAGE. For nucleic acids, urea is the most commonly used denaturant. For proteins, sodium dodecyl sulfate (SDS) is an anionic detergent applied to protein samples to linearize proteins and to impart a negative charge to linearized proteins. 2-Mercaptoethanol may also be used to disrupt the disulfide bonds found in the protein complexes, which helps further linearize the protein. In most proteins, the binding of SDS to the polypeptide chain imparts an even distribution of charge per unit mass, thereby resulting in a fractionation by approximate size during electrophoresis. Proteins that have a greater hydrophobic content, for instance many membrane proteins, and those that interact with surfactants in their native environment, are intrinsically harder to treat accurately using this method, due to the greater variability in the ratio of bound SDS.

Procedure

Sample Preparation

Samples may be any material containing proteins or nucleic acids. These may be biologically derived, for example from prokaryotic or eukaryotic cells, tissues, viruses, environmental samples, or purified proteins. In the case of solid tissues or cells, these are often first broken down mechanically using a blender (for larger sample volumes), using a homogenizer (smaller volumes), by sonicator or by using cycling of high pressure, and a combination of biochemical and mechanical techniques – including various types of filtration and centrifugation – may be used to separate different cell compartments and organelles prior to electrophoresis. Synthetic biomolecules such as oligonucleotides may also be used as analytes.

Reduction of a typical disulfide bond by DTT via two sequential thiol-disulfide exchange reactions.

The sample to analyze is optionally mixed with a chemical denaturant if so desired, usually SDS for proteins or urea for nucleic acids. SDS is an anionic detergent that denatures secondary and non–disulfide–linked tertiary structures, and additionally applies a negative charge to each protein in proportion to its mass. Urea breaks the hydrogen bonds between the base pairs of the nucleic acid, causing the constituent strands to separate. Heating the samples to at least 60 °C further promotes denaturation.

In addition to SDS, proteins may optionally be briefly heated to near boiling in the presence of a reducing agent, such as dithiothreitol (DTT) or 2-mercaptoethanol (beta-mercaptoethanol/BME), which further denatures the proteins by reducing disulfide linkages, thus overcoming some forms of tertiary protein folding, and breaking up quaternary protein structure (oligomeric subunits). This is known as reducing SDS-PAGE.

A tracking dye may be added to the solution. This typically has a higher electrophoretic mobility than the analytes to allow the experimenter to track the progress of the solution through the gel during the electrophoretic run.

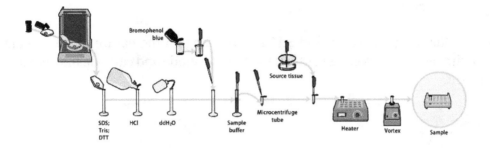

Preparing Acrylamide Gels

The gels typically consist of acrylamide, bisacrylamide, the optional denaturant (SDS or urea), and a buffer with an adjusted pH. The solution may be degassed under a vacuum to prevent the formation of air bubbles during polymerization. Alternatively, butanol may be added to the resolving gel (for proteins) after it is poured, as butanol removes bubbles and makes the surface smooth. A source of free radicals and a stabilizer, such as ammonium persulfate and TEMED are added to initiate polymerization. The polymerization reaction creates a gel because of the added bisacrylamide, which can form cross-links between two acrylamide molecules. The ratio of bisacrylamide to acrylamide can be varied for special purposes, but is generally about 1 part in 35. The acrylamide concentration of the gel can also be varied, generally in the range from 5% to 25%. Lower percentage gels are better for resolving very high molecular weight molecules, while much higher percentages of acrylamide are needed to resolve smaller proteins.

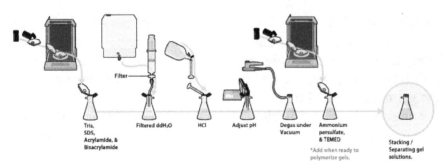

Gels are usually polymerized between two glass plates in a gel caster, with a comb inserted at the top to create the sample wells. After the gel is polymerized the comb can be removed and the gel is ready for electrophoresis.

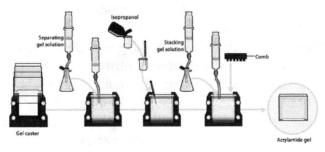

Electrophoresis

Various buffer systems are used in PAGE depending on the nature of the sample and the experimental objective. The buffers used at the anode and cathode may be the same or different.

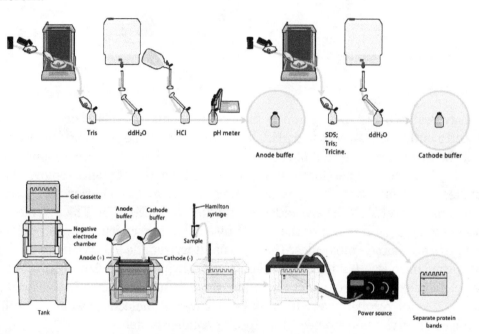

An electric field is applied across the gel, causing the negatively charged proteins or nucleic acids to migrate across the gel away from the negative electrode (which is the cathode being that this is an electrolytic rather than galvanic cell) and towards the positive electrode (the anode). Depending on their size, each biomolecule moves differently through the gel matrix: small molecules more easily fit through the pores in the gel, while larger ones have more difficulty. The gel is run usually for a few hours, though this depends on the voltage applied across the gel; migration occurs more quickly at higher voltages, but these results are typically less accurate than at those at lower voltages. After the set amount of time, the biomolecules have migrated different distances based on their size. Smaller biomolecules travel farther down the gel, while larger ones remain closer to the point of origin. Biomolecules may therefore be separated roughly according to size, which depends mainly on molecular weight under denaturing conditions, but also depends on higher-order conformation under native conditions. The gel mobility is defined as the rate of migration traveled with a voltage gradient of 1V/cm and has units of $cm^2/sec/V$. For analytical purposes, the relative mobility of biomolecules, R_f, the ratio of the distance the molecule traveled on the gel to the total travel distance of a tracking dye is plotted versus the molecular weight of the molecule (or sometimes the log of MW, or rather the M_r, molecular radius). Such typically linear plots represent the standard markers or calibration curves that are widely used for the quantitative estimation of a variety of biomolecular sizes.

Certain glycoproteins, however, behave anomalously on SDS gels. Additionally, the analysis of larger proteins ranging from 250,000 to 600,000 Da is also reported to be problematic due to the fact that such polypeptides move improperly in the normally used gel systems.

Further Processing

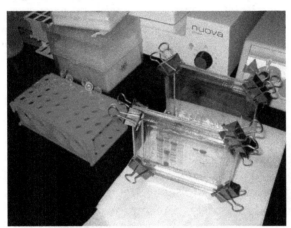

Two SDS-PAGE-gels after a completed run

Following electrophoresis, the gel may be stained (for proteins, most commonly with Coomassie Brilliant Blue R-250 or autoradiography; for nucleic acids, ethidium bromide; or for either, silver stain), allowing visualization of the separated proteins, or processed further (e.g. Western blot). After staining, different species biomolecules appear as distinct bands within the gel. It is common to run molecular weight size markers of known molecular weight in a separate lane in the gel to calibrate the gel and determine the approximate molecular mass of unknown biomolecules by comparing the distance traveled relative to the marker.

For proteins, SDS-PAGE is usually the first choice as an assay of purity due to its reliability and ease. The presence of SDS and the denaturing step make proteins separate, approximately based on size, but aberrant migration of some proteins may occur. Different proteins may also stain differently, which interferes with quantification by staining. PAGE may also be used as a preparative technique for the purification of proteins. For example, quantitative preparative native continuous polyacrylamide gel electrophoresis (QPNC-PAGE) is a method for separating native metalloproteins in complex biological matrices.

Chemical Ingredients and their Roles

Polyacrylamide gel (PAG) had been known as a potential embedding medium for sectioning tissues as early as 1964, and two independent groups employed PAG in electrophoresis in 1959. It possesses several electrophoretically desirable features that make it a versatile medium. It is a synthetic, thermo-stable, transparent, strong, chemically relatively inert gel, and can be prepared with a wide range of average pore sizes. The pore size of a gel and the reproducibilty in gel pore size are determined by three factors, the total amount of acrylamide present (%T) (T = Total concentration of acrylamide and bisacrylamide monomer), the amount of cross-linker (%C) (C = bisacrylamide concentration), and the time of polymerization of acrylamide (cf. QPNC-PAGE). Pore size decreases with increasing %T; with cross-linking, 5%C gives the smallest pore size. Any increase or decrease in %C from 5% increases the pore size, as pore size with respect to %C is a parabolic function with vertex as 5%C. This appears to be because of non-homogeneous bundling of polymer strands within the gel. This gel material can also withstand high voltage gradients, is amenable to various staining and destaining procedures, and can be digested to extract separated fractions or dried for autoradiography and permanent recording.

Components

Polyacrylamide gels are composed of a stacking gel and separating gel. Stacking gels have a higher porosity relative to the separating gel, and allow for proteins to migrate in a concentrated area. Additionally, stacking gels usually have a pH of 6.8, since the neutral glycine molecules allow for faster protein mobility. Separating gels have a pH of 8.8, where the anionic glycine slows down the mobility of proteins. Separating gels allow for the separation of proteins and have a relatively lower porosity. Here, the proteins are separated based on size (in SDS-PAGE) and size/ charge (Native PAGE).

Chemical buffer stabilizes the pH value to the desired value within the gel itself and in the electrophoresis buffer. The choice of buffer also affects the electrophoretic mobility of the buffer counterions and thereby the resolution of the gel. The buffer should also be unreactive and not modify or react with most proteins. Different buffers may be used as cathode and anode buffers, respectively, depending on the application. Multiple pH values may be used within a single gel, for example in DISC electrophoresis. Common buffers in PAGE include Tris, Bis-Tris, or imidazole.

Counterion balance the intrinsic charge of the buffer ion and also affect the electric field strength during electrophoresis. Highly charged and mobile ions are often avoided in SDS-PAGE cathode buffers, but may be included in the gel itself, where it migrates ahead of the protein. In applications such as DISC SDS-PAGE the pH values within the gel may vary to change the average charge of the counterions during the run to improve resolution. Popular counterions are glycine and tricine. Glycine has been used as the source of trailing ion or slow ion because its pKa is 9.69 and mobility of glycinate are such that the effective mobility can be set at a value below that of the slowest known proteins of net negative charge in the pH range. The minimum pH of this range is approximately 8.0.

Acrylamide (C_3H_5NO; mW: 71.08) when dissolved in water, slow, spontaneous autopolymerization of acrylamide takes place, joining molecules together by head on tail fashion to form long single-chain polymers. The presence of a free radical-generating system greatly accelerates polymerization. This kind of reaction is known as Vinyl addition polymerisation. A solution of these polymer chains becomes viscous but does not form a gel, because the chains simply slide over one another. Gel formation requires linking various chains together. Acrylamide is carcinogenic, a neurotoxin, and a reproductive toxin. It is also essential to store acrylamide in a cool dark and dry place to reduce autopolymerisation and hydrolysis.

Bisacrylamide (N,N'-Methylenebisacrylamide) ($C_7H_{10}N_2O_2$; mW: 154.17) is the most frequently used cross linking agent for polyacrylamide gels. Chemically it can be thought of as two acrylamide molecules coupled head to head at their non-reactive ends. Bisacrylamide can crosslink two polyacrylamide chains to one another, thereby resulting in a gel.

Sodium dodecyl sulfate (SDS) ($C_{12}H_{25}NaO_4S$; mW: 288.38) (only used in denaturing protein gels) is a strong detergent agent used to denature native proteins to unfolded, individual polypeptides. When a protein mixture is heated to 100 °C in presence of SDS, the detergent wraps around the polypeptide backbone. It binds to polypeptides in a constant weight ratio of 1.4 g SDS/g of polypeptide. In this process, the intrinsic charges of polypeptides become negligible when compared to the negative charges contributed by SDS. Thus polypeptides after treatment become rod-like structures possessing a uniform charge density, that is same net negative charge per unit weight. The electrophoretic mobilities of these proteins is a linear function of the logarithms of their molecular weights. Without SDS, different proteins with similar molecular weights would migrate differently due to differences in mass-charge ratio, as each protein has an isoelectric point and molecular weight particular to its primary structure. This is known as native PAGE. Adding SDS solves this problem, as it binds to and unfolds the protein, giving a near uniform negative charge along the length of the polypeptide.

Urea ($CO(NH_2)_2$; mW: 60.06) is a chaotropic agent that increases the entropy of the system by interfering with intramolecular interactions mediated by non-covalent forces such as hydrogen bonds and van der Waals forces. Macromolecular structure is dependent on the net effect of these forces, therefore it follows that an increase in chaotropic solutes denatures macromolecules.

Ammonium persulfate (APS) ($N_2H_8S_2O_8$; mW: 228.2) is a source of free radicals and is often used as an initiator for gel formation. An alternative source of free radicals is riboflavin, which generated free radicals in a photochemical reaction.

TEMED (N, N, N', N'-tetramethylethylenediamine) ($C_6H_{16}N_2$; mW: 116.21) stabilizes free radicals and improves polymerization. The rate of polymerisation and the properties of the resulting gel depend on the concentrations of free radicals. Increasing the amount of free radicals results in a decrease in the average polymer chain length, an increase in gel turbidity and a decrease in gel elasticity. Decreasing the amount shows the reverse effect. The lowest catalytic concentrations that allow polymerisation in a reasonable period of time should be used. APS and TEMED are typically used at approximately equimolar concentrations in the range of 1 to 10 mM.

Chemicals for Processing and Visualization

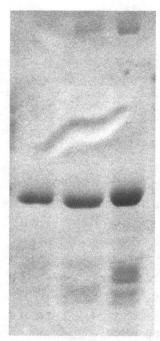

PAGE of rotavirus proteins stained with Coomassie blue

The following chemicals and procedures are used for processing of the gel and the protein samples visualized in it.

Tracking dye: as proteins and nucleic acids are mostly colorless, their progress through the gel during electrophoresis cannot be easily followed. Anionic dyes of a known electrophoretic mobility are therefore usually included in the PAGE sample buffer. A very common tracking dye is Bromophenol blue (BPB, 3',3",5',5" tetrabromophenolsulfonphthalein). This dye is coloured at alkali and neutral pH and is a small negatively charged molecule that moves towards the anode. Being a highly mobile molecule it moves ahead of most proteins. As it reaches the anodic end of the electrophoresis medium electrophoresis is stopped. It can weakly bind to some proteins and impart a blue colour. Other common tracking dyes are xylene cyanol, which has lower mobility, and Orange G, which has a higher mobility.

Loading aids: most PAGE systems are loaded from the top into wells within the gel. To ensure that the sample sinks to the bottom of the gel, sample buffer is supplemented with additives that increase the density of the sample. These additives should be non-ionic and non-reactive towards proteins to avoid interfering with electrophoresis. Common additives are glycerol and sucrose.

Coomassie Brilliant Blue R-250 (CBB)($C_{45}H_{44}N_3NaO_7S_2$; mW: 825.97) is the most popular protein stain. It is an anionic dye, which non-specifically binds to proteins. The structure of CBB is predominantly non-polar, and it is usually used in methanolic solution acidified with acetic acid. Proteins in the gel are fixed by acetic acid and simultaneously stained. The excess dye incorporated into the gel can be removed by destaining with the same solution without the dye. The proteins are detected as blue bands on a clear background. As SDS is also anionic, it may interfere with staining process. Therefore, large volume of staining solution is recommended, at least ten times the volume of the gel.

Ethidium bromide (EtBr) is the traditionally most popular nucleic acid stain.

Silver staining is used when more sensitive method for detection is needed, as classical oomassie Brilliant Blue staining can usually detect a 50 ng protein band, Silver staining increases the sensitivity typically 50 times. The exact chemical mechanism by which this happens is still largely unknown. Silver staining was introduced by Kerenyi and Gallyas as a sensitive procedure to detect trace amounts of proteins in gels. The technique has been extended to the study of other biological macromolecules that have been separated in a variety of supports. Many variables can influence the colour intensity and every protein has its own staining characteristics; clean glassware, pure reagents and water of highest purity are the key points to successful staining. Silver staining was developed in the 14th century for colouring the surface of glass. It has been used extensively for this purpose since the 16th century. The colour produced by the early silver stains ranged between light yellow and an orange-red. Camillo Golgi perfected the silver staining for the study of the nervous system. Golgi's method stains a limited number of cells at random in their entirety.

Autoradiography, also used for protein band detection post gel electrophoresis, uses radioactive isotopes to label proteins, which are then detected by using X-ray film.

Western blotting is a process by which proteins separated in the acrylamide gel are electrophoretically transferred to a stable, manipulable membrane such as a nitrocellulose, nylon, or PVDF membrane. It is then possible to apply immunochemical techniques to visualise the transferred proteins, as well as accurately identify relative increases or decreases of the protein of interest.

Blue Native (BN) PAGE

Blue Native PAGEallows proteins to be studied in their native state and does not involve any denaturation step. The Coomassie dye used herein provides necessary charge to the protein complexes and further helps in their separation during migration on the gel. Thus the electrophoretic mobility mainly relies on the negative charge that is imparted by the dye, and the size and shape of the protein complexes that are being studied. It can be used for identification of multi-protein complexes and hence provides an integrative view of the protein, as they need to be separated in the native conditions.

1DE can be effectively used to determine the whole protein molecular weight, the molecular weight of individual components of a complex protein, detection of isoforms of a single protein and post-translational modifications. It can also be used to purify a particular protein from a complex mixture, which has to be further used for different applications and can provide an integrative view of the protein function.

Two-dimensional Gel Electrophoresis

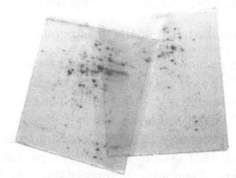

2D-Gels (Coomassie stained)

Two-dimensional gel electrophoresis, abbreviated as 2-DE or 2-D electrophoresis, is a form of gel electrophoresis commonly used to analyze proteins. Mixtures of proteins are separated by two properties in two dimensions on 2D gels. 2-DE was first independently introduced by O'Farrell and Klose in 1975.

Robots are used for the isolation of protein spots from 2D gels in modern laboratories

Basis for Separation

2-D electrophoresis begins with electrophoresis in the first dimension and then separates the molecules perpendicularly from the first to create an electropherogram in the second dimension. In electrophoresis in the first dimension, molecules are separated linearly according to their isoelectric point. In the second dimension, the molecules are then separated at 90 degrees from the first electropherogram according to molecular mass. Since it is unlikely that two molecules will be similar in two distinct properties, molecules are more effectively separated in 2-D electrophoresis than in 1-D electrophoresis.

The two dimensions that proteins are separated into using this technique can be isoelectric point, protein complex mass in the native state, and protein mass.

Separation of the proteins by isoelectric point is called isoelectric focusing (IEF). Thereby, a gradient of pH is applied to a gel and an electric potential is applied across the gel, making one end more positive than the other. At all pH values other than their isoelectric point, proteins will be charged. If they are positively charged, they will be pulled towards the more negative end of the gel and if they are negatively charged they will be pulled to the more positive end of the gel. The proteins applied in the first dimension will move along the gel and will accumulate at their isoelectric point; that is, the point at which the overall charge on the protein is 0 (a neutral charge).

For the analysis of the functioning of proteins in a cell, the knowledge of their cooperation is essential. Most often proteins act together in complexes to be fully functional. The analysis of this sub organelle organisation of the cell requires techniques conserving the native state of the protein complexes. In native polyacrylamide gel electrophoresis (native PAGE), proteins remain in their native state and are separated in the electric field following their mass and the mass of their complexes respectively. To obtain a separation by size and not by net charge, as in IEF, an additional charge is transferred to the proteins by the use of Coomassie Brilliant Blue or lithium dodecyl sulfate. After completion of the first dimension the complexes are destroyed by applying the denaturing SDS-PAGE in the second dimension, where the proteins of which the complexes are composed of are separated by their mass.

Before separating the proteins by mass, they are treated with sodium dodecyl sulfate (SDS) along with other reagents (SDS-PAGE in 1-D). This denatures the proteins (that is, it unfolds them into long, straight molecules) and binds a number of SDS molecules roughly proportional to the protein's length. Because a protein's length (when unfolded) is roughly proportional to its mass, this is equivalent to saying that it attaches a number of SDS molecules roughly proportional to the protein's mass. Since the SDS molecules are negatively charged, the result of this is that all of the proteins will have approximately the same mass-to-charge ratio as each other. In addition, proteins will not migrate when they have no charge (a result of the isoelectric focusing step) therefore the coating of the protein in SDS (negatively charged) allows migration of the proteins in the second dimension (SDS-PAGE, it is not compatible for use in the first dimension as it is charged and a nonionic or zwitterionic detergent needs to be used). In the second dimension, an electric potential is again applied, but at a 90 degree angle from the first field. The proteins will be attracted to the more positive side of the gel (because SDS is negatively charged) proportionally to their mass-to-charge ratio. As previously explained, this ratio will be nearly the same for all proteins. The proteins' progress will be slowed by frictional forces. The gel therefore acts like a molecular sieve when the current is applied, separating the proteins on the basis of their molecular weight with larger proteins being retained higher in the gel and smaller proteins being able to pass through the sieve and reach lower regions of the gel.

Detecting Proteins

The result of this is a gel with proteins spread out on its surface. These proteins can then be detected by a variety of means, but the most commonly used stains are silver and Coomassie Brilliant Blue staining. In the former case, a silver colloid is applied to the gel. The silver binds to cysteine groups within the protein. The silver is darkened by exposure to ultra-violet light. The amount of silver can be related to the darkness, and therefore the amount of protein at a given location on the gel. This measurement can only give approximate amounts, but is adequate for most purposes. Silver staining is 100x more sensitive than Coomassie Brilliant Blue with a 40-fold range of linearity.

Molecules other than proteins can be separated by 2D electrophoresis. In supercoiling assays, coiled DNA is separated in the first dimension and denatured by a DNA intercalator (such as ethidium bromide or the less carcinogenic chloroquine) in the second. This is comparable to the combination of native PAGE /SDS-PAGE in protein separation.

Common Techniques

IPG-DALT

A common technique is to use an Immobilized pH gradient (IPG) in the first dimension. This technique is referred to as IPG-DALT. The sample is first separated onto IPG gel (which is commercially available) then the gel is cut into slices for each sample which is then equilibrated in SDS-mercaptoethanol and applied to an SDS-PAGE gel for resolution in the second dimension. Typically IPG-DALT is not used for quantification of proteins due to the loss of low molecular weight components during the transfer to the SDS-PAGE gel.

2D Gel Analysis Software

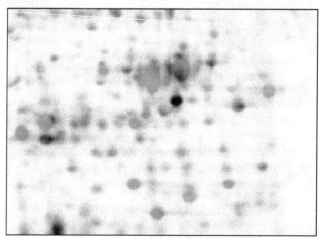

Warping: Images of two 2D electrophoresis gels, overlaid with Delta2D. First image is colored in orange, second one colored in blue. Due to running differences, corresponding spots do not overlap

Electrophoresis: An Integrated Study

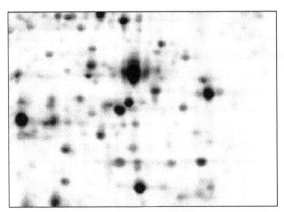

Warping: Images of two 2D electrophoresis gels after warping. First image is colored in orange, second one colored in blue. Corresponding spots overlap after warping. Common spots are colored black, orange spots are only present (or much stronger) on the first image, blue spots are only present (or much stronger) on the second image

In quantitative proteomics, these tools primarily analyze bio-markers by quantifying individual proteins, and showing the separation between one or more protein "spots" on a scanned image of a 2-DE gel. Additionally, these tools match spots between gels of similar samples to show, for example, proteomic differences between early and advanced stages of an illness. Software packages include Delta2D, ImageMaster, Melanie, PDQuest, Progenesis and REDFIN – among others. While this technology is widely utilized, the intelligence has not been perfected. For example, while PDQuest and Progenesis tend to agree on the quantification and analysis of well-defined well-separated protein spots, they deliver different results and analysis tendencies with less-defined less-separated spots.

Challenges for automatic software-based analysis include incompletely separated (overlapping) spots (less-defined and/or separated), weak spots / noise (e.g., "ghost spots"), running differences between gels (e.g., protein migrates to different positions on different gels), unmatched/undetected spots, leading to missing values, mismatched spots , errors in quantification (several distinct spots may be erroneously detected as a single spot by the software and/or parts of a spot may be excluded from quantification), and differences in software algorithms and therefore analysis tendencies.

Generated picking lists can be used for the automated in-gel digestion of protein spots, and subsequent identification of the proteins by mass spectrometry.

For an overview of the current approach for software analysis of 2DE gel images.

Staining Methods

Different staining methods are commonly used to stain the 1D or 2D gels.

1. Coomassie Brilliant Blue (CBB): Coomassie blue dyes (R-250 & G-250) are low cost, organic dyes that are easy to use for staining of proteins that have been

separated by electrophoretic techniques. Gels are soaked in the dye solution dissolved in methanol and acetic acid, after which excess stain is then washed off with a destaining solution. The higher affinity of proteins towards the dye molecules, allows the protein bands to be selectively stained with sensitivity of 8-100 ng without significant staining of the background. These dyes are also compatible for further MS-based applications.

2. Silver staining: Electrophoresis gels are saturated with silver ions in the form of either silver nitrate or as an ammonia-silver complex after fixing the proteins in the gel. The less tightly bound metal ions are subsequently washed off and the protein-bound silver ions are reduced to metallic silver using formaldehyde under alkaline conditions in presence of sodium carbonate or citrate buffer solution. Although as little as 1 ng of protein can be detected by silver staining, the gel-to-gel reproducibility remains an issue. Compatibility of silver stains with MS is another issue, which has however, been overcome in recently introduced silver stains.

3. SYPRO Ruby Red: This is a ruthenium-based metal chelate fluorescent stain that provides a single step protein staining procedure with low background staining in polyacrylamide gels. They have been observed to be as sensitive as the silver stains (0.25-1 ng) with the liner dynamic range extending over three orders of magnitude, thereby showing better performance than CBB and silver stains. This stain can also be combined with other dyes thereby allowing multiple detections in a single gel.

4. SYPRO orange: This dye is less sensitive than SYPRO Ruby Red but is also capable of detecting proteins in SDS-PAGE gels in a rapid single step process without the requirement for any destaining procedure. As little as 4-30 ng of protein can be detected by this fluorescent dye and it is compatible for further MS-based applications. Two other similar dyes having comparable sensitivities and similar excitation and emission wavelengths are SYPRO Red and Tangerine.

5. Cyanine dyes: They are water-soluble derivatives of N-hydroxy succinimide that covalently bind the e-amino groups of a protein's lysine residues and are spectrally resolvable as they fluoresce at distinct wavelengths. The labeled protein samples can therefore be mixed and run on a single gel, thereby eliminating the problem of gel-to-gel variations, the principle employed in difference gel electrophoresis (DIGE). Cy3, Cy5 and Cy2 having sensitivities of 0.1-2 ng are most commonly used for proteomic and MS-based applications.

6. Lightning fast/Deep purple: A fluorescence-based stain obtained from the fungus Epicoccum nigrum that can be used for detecting proteins in 1-D and 2-D gel electrophoresis with sensitivity down to 100 pg protein. Stained proteins are excited by near-UV or visible light with maximum fluorescence emission

occurring at around 610nm. These dyes are suited for further use with Edman or MS applications.

7. Pro-Q-Diamond: This fluorescent dye is capable of detecting modified proteins that have been phosphorylated at their serine, threonine or tyrosine residues. They are suitable for use with electrophoretic techniques or with protein microarrays and offer sensitivity down to few ng levels, depending upon the format in which they are used. This dye can also be combined with other staining procedures thereby allowing more than one detection protocol on a single gel.

Fluorescence Based Difference in Gel Electrophoresis (DIGE)

There are various limitations associated with 2-DE due to the gel-to-gel variations and manual artifacts, which emerge mainly from inconsistency in sample preparation, and then from the subsequent gel running itself, during the 1st and 2nd dimensions. These limitations eventually lead to the lack of reproducibility. This has necessitated the need for the development of a technique, which would overcome these limitations and help in solving the purpose of protein studies in a better way.

Comparison of Techniques

TECHNIQUE	ADVANTAGES	DISADVANTAGES	APPLICATIONS
1DE	• Easy technique to check for protein purity in a given protein extraction.	• Large number of proteins cannot be separated from a complex mixture with good resolution. • It cannot be used to study whole proteome or to analyze complex fluids like serum or cell lysates.	• Test protein purity • Study protein-expression
2DE	• Powerful technique for simultaneous separation of thousands of proteins. • Highly sensitive visualization of proteins as small differences in protein expression levels can be detected with statistical confidence. • Relatively easy to handle and affordable	• Laborious and time consuming. • Requires 800-1000 ug of protein as the starting material. • In one gel, only one sample can be analyzed. • Requires procedures like staining or fixation to be done after the second dimension gel electrophoresis	• Study global and differential protein expression and resolution of complex proteins. (Chen et al. 2004) • For Biomarker Discovery (Lescuyer et al. 2007)

DIGE	• Requires less amount of starting protein material. • Highly reproducible and sensitive as two different samples can be analyzed on the same gel and so the differences in protein expression levels are purely attributed to biological variations. • Better quantitative comparison is attributed to the use of internal control. • No need for fixation or destaining as fluorescent dyes are used for sample labeling.	• Only two different types of samples can be analyzed on a single gel. • Spot excision is a problem since they are not visible to the naked eye and hence require aids such as robotic Spot picking. • Fluorescence detection gives high background at times when signals from different labels may get mixed.	• In Cancer studies (Bai et al. 2010) • For Biomarker discovery (Uemura et al. 2009) • For studying post translational modifications (DeKroon et al. 2012)

Difference in Gel Electrophoresis

Proteome studies carried out by the traditional 2DE technique have limitations with respect to reproducibility, which can be attributed to gel-to-gel variations. These variations affect the quantitative comparison of protein expression levels. To address these issues and enhance the reproducibility, advanced gel-based technique, namely, Difference in Gel Electrophoresis was reported by Unlu et al. in 1997. This technique exploits the fact that two different proteins when labeled with two different fluorescent dyes (Cy3 and Cy5) can be visualized individually on one single gel. These fluorescent dyes are pH insensitive, photo-stable and spectrally distinct. Moreover, the use of an internal control, which is a pool of control and test sample, is labeled with a third fluorescent dye (Cy2), to facilitate co-detection, normalization and accurate quantification of protein samples.

Box-1: Abbreviations & Terminologies

- DIGE: Difference In Gel Electrophoresis.

- Cy Dyes: Cyanine dyes used for labeling of protein samples from different sources, which can be mixed and run together using electrophoresis process.

- Internal Control: A pool of equal amount of the two samples under study, which is labeled with a third fluorescent dye (usually Cy2).

- Codetection: Simultaneous detection of protein spots from two different spots, which is enabled due to the multiplexing ability of DIGE.

- DeCyder™: DIGE Image Analysis software
- DIA: Differential In-gel Analysis module
- BVA: Biological Variation Analysis module

Workflow for DIGE

- Sample Preparation & Labeling
- IPG Strip Rehydration
- First dimension (IEF)
- Second dimension (SDS-PAGE)
- Scanning
- Image analysis and Data interpretation
- Protein identification of significant spots

Sample Preparation and Labeling

The protein pellet dissolved in appropriate buffer is taken as the starting sample for labeling. The sample pH should be 8.5, and can be adjusted using 100mM NaOH. Each of the samples is labeled with one of the fluorescent dyes (for e.g. the control sample can be labeled with Cy3 and the test sample can be labeled with Cy5, or vice versa). An internal pool containing equal amount of the control and treated samples is mixed together and is labeled with a different fluorescent dye, usually Cy2. These dyes bind to the ε-amine groups of the protein's lysine residues. The dye swapping should be done to check reproducibility and efficiency of the working protocol. The reaction is quenched using 10 mM lysine, which combines with unbound dye molecules and stops the reaction. The labeled samples are then stored at 4°C, until they are further rehydrated.

Cyanine Dyes

- N-hydroxy succinimidyl ester derivatives of cyanine.
- Bind to ε-amine groups of the protein's lysine residues.
- Lysine is targeted for cyanine labeling.
- Spectrally resolvable fluors that are matched for mass and charge.
- There is no change in signal over wide pH range used during first-dimension (IEF) separation.

- Discrete signal from each fluor with minimal cross talk contributes to high accuracy.

- Absorbance max for Cy3 dye is 550 nm and emission max 570 nm; and absorbance max for Cy5 dye is 649 nm and emission max 670 nm.

Rehydration of IPG Strips and IEF

The IPG strips need to be rehydrated with the labeled sample, which is mixed with the respective IPG buffer. The sample is then placed over the IPG strip and rehydration is allowed to occur by incubating the strips at RT for 16 hrs. After 1 hr, the samples are overlaid with mineral oil to prevent sample evaporation. The rehydrated strips are further used for isoelectric focusing (IEF). In IEF, the rehydrated strips are first placed in the IEF tray and program for IEF run is selected as per the strip length and pH range. In IEF, the applied voltage gradient brings about separation of proteins on the basis of their isoelectric point (pI), at which the net charge on a given protein is zero.

Second Dimension (SDS-PAGE)

The separation of proteins on the basis of their molecular weight is brought about in the Second dimension by SDS-PAGE. The focused strip is run on the gel in a direction, which is orthogonal to the IEF run.

Scanning and Image Acquisition

After the second dimension run is completed, the gels are scanned at 3 different wavelengths, which are the excitation wavelengths of each of the 3 dyes, namely Cy3, Cy5 and Cy2. The resulting images are stored for further analysis.

Image Analysis and Data Interpretation

The saved images are further analyzed by using software especially designed for 2D-DIGE, such as DeCyder. The pair-wise analysis between test and control samples can be performed using Differential In-gel Analysis (DIA) module, whereas the analysis between multiple samples belonging to two different groups can be performed using the Biological Variation Analysis (BVA) module of the software.

Applications of DIGE

The 2D-DIGE technique has been used for several proteomic applications. Few representative studies are described here.

DIGE for detection of markers for cancer:

- Human tissue biopsies are obtained and protein is extracted. The 2D-DIGE is carried out as per the described protocol. The control samples of healthy control labeled with Cy3, whereas the treated samples (diseased patient for cancers) labeled with Cy5 or vice versa. An internal pool is labeled with Cy2.

APPLICATION	REFERENCE
Proteome analysis of human colorectal cancer tissue using DIGE	Xui bai et al. 2010
Biomarker discovery for esophageal cancer	Norihisa Uemura et al. 2009
Whole proteome analysis of drug treated rat muscle	Kenyani et al. 2011
Post-translational modifications	DeKroon RM et al. 2012

Advantages and Limitations of DIGE

Advantages

- The amount of protein starting material required is very small as compared to the requirement for 2DE. DIGE is extremely sensitive, i.e. <1 fmol of protein can be detected and it can also enable the linear detection over a >10,000-fold protein abundance range.

- Higher reproducibility as two different samples can be analyzed on the same gel; therefore, differences in protein expression levels are purely attributed to biological variations.

- Accurate quantitative comparison, which can be attributed to the use of internal control, which allows normalization of spot intensities across different, gels and increases accuracy of protein expression differences.

- No post-electrophoretic processing (fixation or destaining) is necessary and thereby there is a reduction in protein loss particularly in the low molecular weight range.

- User bias can be eliminated by the co-detection method.

Limitations

- Only two different samples can be analyzed on a single gel. For more than two samples, a large number of pair-wise comparisons would be required. Mass spectrometry based techniques such as iTRAQ could be used for 4 or 8-plexing.

- Spot excision is a problem since spots are not visible to the naked eye and hence require aids such as Robotic Spot picking. Alternatively, a normal 2-DE gel has to be run for these samples, which is then stained with Coomassie dye and spots of interest are excised.

- Fluorescence detection comes with several inherent problems such as high background, the detection of signals from non-protein sources (e.g. dust and residue on plates) and overlap of signals from different fluorophores.

DIGE technique has gained wide acceptance in proteomic studies because of its sensitive quantification abilities and reduced analytical variability. It also offers an advantage over 2-DE with respect to the reproducibility and accuracy in quantification. DIGE is an increasingly employed proteomic technique, which can be used to address several biological questions.

Ion Chromatography

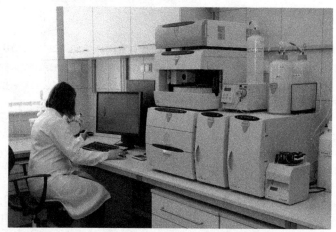

Thermo Scientific Dionex ICS-5000⁺ ion chromatography system

Ion chromatography (or ion-exchange chromatography) is a chromatography process that separates ions and polar molecules based on their affinity to the ion exchanger. It works on almost any kind of charged molecule—including large proteins, small nucleotides, and amino acids. It is often used in protein purification, water analysis, and quality control. The water-soluble and charged molecules such as proteins, amino acids, and peptides bind to moieties which are oppositely charged by forming ionic bonds to the insoluble stationary phase. The equilibrated stationary phase consists of an ionizable functional group where the targeted molecules of a mixture to be separated and quantified can bind while passing through the column—a cationic stationary phase is used to separate anions and an anionic stationary phase is used to separate cations. Cation exchange chromatography is used when the desired molecules to separate are cations and anion exchange chromatography is used to separate anions. The bound molecules then can be eluted and collected using an eluant which contains anions and cations by running higher concentration of ions through the column or changing pH of the column. One of the primary advantages for the use of ion chromatography is only one interaction involved during the separation as opposed to other separation techniques; therefore, ion chromatography may have higher matrix tolerance. However, there are also disadvantages involved when performing ion-exchange chromatography, such as constant evolution with the technique which leads to the inconsistency from column to column.

History

Table: History of ion exchange and ion chromatography, the analytical technique based on ion exchange

Date	Development	Researchers
ca. 1850	Soil as an ion exchanger for Mg2+, Ca2+ and NH4+	Thomson & Way
1935	Sulfonated and aminated condensation polymers (phenol/formaldehyde)	Adams, Homes
1942	Sulfonated PS/DVB resin as cation exchanger (Manhattan Project)	d'Alelio
1947	Aminated PS/DVB resin as anion exchanger	McBurney
1953	Ion exclusion chromatography	Wheaton, Baumann
1957	Macroporous ion exchangers	Corte, Meyer, Kunin et al.
1959	Basic Theoretical principles	Helfferich
1967-70	Pellicular ion exchnagers	Horvath, Kirkland
1975	Ion exchange chromatography with conductivity detection using a "stripper"	Small, Stevens, Bauann
1979	Conductivity detection without a "stripper"	Gjerde, Fritz, Schmuckler
1976-80	Ion pair chromatography	Waters, Bidlingmeier, Horvath et al.

Ion exchange chromatography

Ion chromatography has advanced through the accumulation of knowledge over a course of many years. Starting from 1947, Spedding and Powell used displacement ion-exchange chromatography for the separation of the rare earths. Additionally, they showed the ion-exchange separation of 14N and 15N isotopes in ammonia. At the start of the 1950s, Kraus and Nelson demonstrated the use of many analytical methods for metal ions dependent on their separation of their chloride, fluoride, nitrate or sulfate complexes by anion chromatography. Automatic in-line detection was progressively introduced from 1960 to 1980 as well as novel chromatographic methods for metal ion separations. A groundbreaking method by Small, Stevens and Bauman at Dow Chemical Co. unfolded the creation of the modern ion chromatography. Anions and cations could now be separated efficiently by a system of suppressed conductivity detection. In 1979, a method for anion chromatography with non-suppressed conductivity detection was introduced by Gjerde et al. Following it in 1980, was a similar method for cation chromatography.

As a result, a period of extreme competition began within the IC market, with supporters for both suppressed and nonsuppressed conductivity detection. This competition led to fast growth of new forms and the fast evolution of IC. A challenge that needs to be overcome in the future development of IC is the preparation of highly efficient monolithic ion-exchange columns and overcoming this challenge would be of great importance to the development of IC.

The boom of Ion exchange chromatography primarily began between 1935–1950 during World War II and it was through the "Manhattan project" that applications and IC were significantly extended. Ion chromatography was originally introduced by two English researchers, agricultural Sir Thompson and chemist J T Way. The works of Thompson and Way involved the action of water-soluble fertilizer salts, ammonium sulfate and potassium chloride. These salts could not easily be extracted from the ground due to the rain. They performed ion methods to treat clays with the salts, resulting in the extraction of ammonia in addition to the release of calcium. It was in the fifties and sixties that theoretical models were developed for IC for further understanding and it was not until the seventies that continuous detectors were utilized, paving the path for the development from low-pressure to high-performance chromatography. Not until 1975 was "ion chromatography" established as a name in reference to the techniques, and was thereafter used as a name for marketing purposes. Today IC is important for investigating aqueous systems, such as drinking water. It is a popular method for analyzing anionic elements or complexes that help solve environmentally relevant problems. Likewise, it also has great uses in the semiconductor industry.

Because of the abundant separating columns, elution systems, and detectors available, chromatography has developed into the main method for ion analysis.

When this technique was initially developed, it was primarily used for water treatment. Since 1935, ion exchange chromatography rapidly manifested into one of the most heavily leveraged techniques, with its principles often being applied to majority of fields of chemistry, including distillation, adsorption, and filtration.

Principle

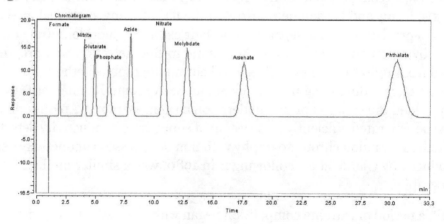

Ion Chromatography

Ion-exchange chromatography separates molecules based on their respective charged groups. Ion-exchange chromatography retains analyte molecules on the column based on coulombic (ionic) interactions. Essentially, molecules undergo electrostatic interactions with opposite charges on the stationary phase matrix. The stationary phase consists of an immobile matrix that contains charged ionizable functional groups or ligands. The stationary phase surface displays ionic functional groups (R-X) that interact with analyte ions of opposite charge. To achieve electroneutrality, these inert charges couple with exchangeable counterions in the solution. Ionizable molecules that are to be purified compete with these exchangeable counterions for binding to the immobilized charges on the stationary phase. These ionizable molecules are retained or eluted based on their charge. Initially, molecules that do not bind or bind weakly to the stationary phase are first to wash away. Altered conditions are needed for the elution of the molecules that bind to the stationary phase. The concentration of the exchangeable counterions, which competes with the molecules for binding, can be increased or the pH can be changed. A change in pH affects the charge on the particular molecules and, therefore, alters binding. The molecules then start eluting out based on the changes in their charges from the adjustments. Further such adjustments can be used to release the protein of interest. Additionally, concentration of counterions can be gradually varied to separate ionized molecules. This type of elution is called gradient elution. On the other hand, step elution can be used in which the concentration of counterions are varied in one step. This type of chromatography is further subdivided into cation exchange chromatography and anion-exchange chromatography. Positively charged molecules bind to anion exchange resins while negatively charged molecules bind to cation exchange resins. The ionic compound consisting of the cationic species M+ and the anionic species B- can be retained by the stationary phase.

Cation exchange chromatography retains positively charged cations because the stationary phase displays a negatively charged functional group:

$$R\text{-}X^-C^+ + M^+B^- \rightleftharpoons R\text{-}X^-M^+ + C^+ + B^-$$

Anion exchange chromatography retains anions using positively charged functional group:

$$R\text{-}X^+A^- + M^+B^- \rightleftharpoons R\text{-}X^+B^- + M^+ + A^-$$

Note that the ion strength of either C^+ or A^- in the mobile phase can be adjusted to shift the equilibrium position, thus retention time.

The ion chromatogram shows a typical chromatogram obtained with an anion exchange column.

Procedure

It is possible to perform ion exchange chromatography in bulk, on thin layers of medium such as glass or plastic plates coated with a layer of the desired stationary phase, or in chromatography columns. Thin layer chromatography or column chromatography share similarities in that they both act within the same governing principles; there is constant and frequent exchange of molecules as the mobile phase travels along the stationary phase. It is not imperative to add the sample in minute volumes as the predetermined conditions for the exchange column have been chosen so that there will be strong interaction between the mobile and stationary phases. Furthermore, the mechanism of the elution process will cause a compartmentalization of the differing molecules based on their respective chemical characteristics. This phenomenon is due to an increase in salt concentrations at or near the top of the column, thereby displacing the molecules at that position, while molecules bound lower are released at a later point when the higher salt concentration reaches that area. These principles are the reasons that ion exchange chromatography is an excellent candidate for initial chromatography steps in a complex purification procedure as it can quickly yield small volumes of target molecules regardless of a greater starting volume.

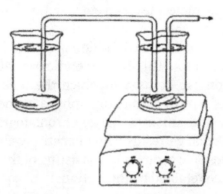

Chamber (left) contains high salt concentration. Stirred chamber (right) contains low salt concentration. Gradual stirring causes the formation of a salt gradient as salt travel from high to low concentrations

Comparatively simple devices are often used to apply counterions of increasing gradient to a chromatography column. Counterions such as Copper(II) are chosen most often for effectively separating peptides and amino acids through complex formation.

A simple device can be used to create a salt gradient. Elution buffer is consistently being drawn from the chamber into the mixing chamber, thereby altering its buffer concentration. Generally, the buffer placed into the chamber is usually of high initial concentration, whereas the buffer placed into the stirred chamber is usually of low concentration. As the high concentration buffer from the left chamber is mixed and drawn into the column, the buffer concentration of the stirred column gradually increase. Altering the shapes of the stirred chamber, as well as of the limit buffer, allows for the production of concave, linear, or convex gradients of counterion.

A multitude of different mediums are used for the stationary phase. Among the most common immobilized charged groups used are trimethylaminoethyl (TAM), triethylaminoethyl (TEAE), diethyl-2-hydroxypropylaminoethyl (QAE), aminoethyl (AE), diethylaminoethyl (DEAE), sulpho (S), sulphomethyl (SM), sulphopropyl (SP), carboxy (C), and carboxymethyl (CM).

Successful column packing is an important aspect of ion chromatography. Stability and efficiency of a final column depends on packing methods, solvent used, and factors that affect mechanical properties of the column. In contrast to early inefficient dry- packing methods, wet slurry packing, in which particles that are suspended in an appropriate solvent are delivered into a column under pressure, shows significant improvement. Three different approaches can be employed in performing wet slurry packing: the balanced density method (solvent's density is about that of porous silica particles), the high viscosity method (a solvent of high viscosity is used), and the low viscosity slurry method (performed with low viscosity solvents).

Polystyrene is used as a medium for ion- exchange. It is made from the polymerization of styrene with the use of divinylbenzene and benzoyl peroxide. Such exchangers form hydrophobic interactions with proteins which can be irreversible. Due to this property, polystyrene ion exchangers are not suitable for protein separation. They are used on the other hand for the separation of small molecules in amino acid separation and removal of salt from water. Polystyrene ion exchangers with large pores can be used for the separation of protein but must be coated with a hydrophillic substance.

Cellulose based medium can be used for the separation of large molecules as they contain large pores. Protein binding in this medium is high and has low hydrophobic character. DEAE is an anion exchange matrix that is produced from a positive side group of diethylaminoethyl bound to cellulose or Sephadex.

Agarose gel based medium contain large pores as well but their substitution ability is lower in comparison to dextrans. The ability of the medium to swell in liquid is based on the cross-linking of these substances, the pH and the ion concentrations of the buffers used.

Incorporation of high temperature and pressure allows a significant increase in the efficiency of ion chromatography, along with a decrease in time. Temperature has an influence of selectivity due to its effects on retention properties. The retention factor ($k = (t_R^g - t_M^g)/(t_M^g - t_{ext})$) increases with temperature for small ions, and the opposite trend is observed for larger ions.

Despite ion selectivity in different mediums, further research is being done to perform ion exchange chromatography through the range of 40–175 °C.

An appropriate solvent can be chosen based on observations of how column particles behave in a solvent. Using an optical microscope, one can easily distinguish a desirable dispersed state of slurry from aggregated particles.

Weak and Strong Ion Exchangers

A "strong" ion exchanger will not lose the charge on its matrix once the column is equilibrated and so a wide range of pH buffers can be used. "Weak" ion exchangers have a range of pH values in which they will maintain their charge. If the pH of the buffer used for a weak ion exchange column goes out of the capacity range of the matrix, the column will lose its charge distribution and the molecule of interest may be lost. Despite the smaller pH range of weak ion exchangers, they are often used over strong ion exchangers due to their having greater specificity. In some experiments, the retention times of weak ion exchangers are just long enough to obtain desired data at a high specificity.

Resins of ion exchange columns may include functional groups such as weak/strong acids and weak/strong bases. There are also special columns that have resins with amphoteric functional groups that can exchange both cations and anions. Examples of functional groups of strong ion exchange resins are quaternary ammonium (Q), which is an anion exchanger, and sulfonic acid (S), which is a cation exchanger. These types of exchangers can maintain their charge density over a pH range of 0–14. Examples of functional groups of Weak ion exchange resins include diethylaminoethyl (DEAE), which is an anion exchanger, and carboxymethyl (CM), which is a cation exchanger. These two types of exchangers can maintain the charge density of their columns over a pH range of 5–9.

In ion chromatography, the interaction of the solute ions and the stationary phase based on their charges determines which ions will bind and to what degree. When the stationary phase features positive groups which attracts anions, it is called an anion exchanger; when there are negative groups on the stationary phase, cations are attracted and it is a cation exchanger. The attraction between ions and stationary phase also depends on the resin, organic particles used as ion exchangers.

Each resin features relative selectivity which varies based on the solute ions present who will compete to bind to the resin group on the stationary phase. The selectivity coefficient, the equivalent to the equilibrium constant, is determined via a ratio of the concentrations between the resin and each ion, however, the general trend is that ion exchangers prefer binding to the ion with a higher charge, smaller hydrated radius, and higher polarizability, or the ability for the electron cloud of an ion to be disrupted by other charges. Despite this selectivity, excess amounts of an ion with a lower selectivity introduced to the column would cause the lesser ion to bind more to the stationary phase as the selectivity coefficient allows fluctuations in the binding reaction that takes place during ion exchange chromatography.

Typical Technique

Metrohm 850 Ion chromatography system

A sample is introduced, either manually or with an autosampler, into a sample loop of known volume. A buffered aqueous solution known as the mobile phase carries the sample from the loop onto a column that contains some form of stationary phase material. This is typically a resin or gel matrix consisting of agarose or cellulose beads with covalently bonded charged functional groups. Equilibration of the stationary phase is needed in order to obtain the desired charge of the column. If the column is not properly equilibrated the desired molecule may not bind strongly to the column. The target analytes (anions or cations) are retained on the stationary phase but can be eluted by increasing the concentration of a similarly charged species that displaces the analyte ions from the stationary phase. For example, in cation exchange chromatography, the positively charged analyte can be displaced by adding positively charged sodium ions. The analytes of interest must then be detected by some means, typically by conductivity or UV/visible light absorbance.

Control an IC system usually requires a chromatography data system (CDS). In addition to IC systems, some of these CDSs can also control gas chromatography (GC) and HPLC.

Membrane Exchange Chromatography

A type of ion exchange chromatography, membrane exchange is a relatively new method of purification designed to overcome limitations of using columns packed with beads. Membrane Chromatographic devices are cheap to mass-produce and disposable unlike other chromatography devices that require maintenance and time to revalidate. There are three types of membrane absorbers that are typically used when separating substances. The three types are flat sheet, hollow fibre, and radial flow. The most common absorber and best suited for membrane chromatography is multiple flat sheets because it has more absorbent volume. It can be used to overcome mass transfer limitations and pressure drop, making it especially advantageous for isolating and purifying viruses, plasmid DNA, and other large macromolecules. The column is packed with microporous membranes with internal pores which contain adsorptive moieties that can bind the target protein. Adsorptive membranes are available in a variety of geometries and chemistry which allows them to be used for purification and also fractionation, concentration, and clarification in an efficiency that is 10 fold that of using beads. Membranes can be prepared through isolation of the membrane itself, where membranes are cut into squares and immobilized. A more recent method involved the use of live cells that are attached to a support membrane and are used for identification and clarification of signaling molecules.

Separating Proteins

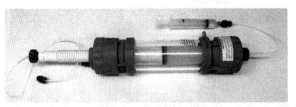

Preparative-scale ion exchange column used for protein purification

Ion exchange chromatography can be used to separate proteins because they contain charged functional groups. The ions of interest (in this case charged proteins) are exchanged for another ions (usually H^+) on a charged solid support. The solutes are most commonly in a liquid phase, which tends to be water. Take for example proteins in water, which would be a liquid phase that is passed through a column. The column is commonly known as the solid phase since it is filled with porous synthetic particles that are of a particular charge. These porous particles are also referred to as beads, may be aminated (containing amino groups) or have metal ions in order to have a charge. The column can be prepared using porous polymers, for macromolecules over 100,000 the optimum size of the porous particle is about 1 µm². This is because slow diffusion of the solutes within the pores does not restrict the separation quality. The beads containing positively charged groups, which attract the negatively charged proteins, are commonly referred to as anion exchange resins. The amino acids that have negatively charged side chains at pH 7 (pH of water) are glutamate and aspartate. The beads that are negatively charged are called cation exchange resins, as positively charged proteins will be attracted. The amino acids that have positively charged side chains at pH 7 are lysine, histidine and asparagine.

The isoelectric point is the pH at which a compound - in this case a protein - has no net charge. A protein's isoelectric point or PI can be determined using the pKa of the side chains, if the amino (positive chain) is able to cancel out the carboxyl (negative) chain, the protein would be at its PI. Using buffers instead of water for proteins that do not have a charge at pH 7, is a good idea as it enables the manipulation of pH to alter ionic interactions between the proteins and the beads. Weakly acid or basic side chains (such as in leucine, proline, alanine, valine, glycine...etc.) are able to have a charge if the pH is high enough to deprotonate the amino group. Separation can be achieved based on the natural isoelectric point of the protein. Alternatively a peptide tag can be genetically added to the protein to give the protein an isoelectric point away from most natural proteins (e.g., 6 arginines for binding to a cation-exchange resin or 6 glutamates for binding to an anion-exchange resin such as DEAE-Sepharose).

Elution by increasing ionic strength of the mobile phase is more subtle. It works because ions from the mobile phase interact with the immobilized ions on the stationary phase, thus "shielding" the stationary phase from the protein, and letting the protein elute.

Elution from ion-exchange columns can be sensitive to changes of a single charge- chromatofocusing. Ion-exchange chromatography is also useful in the isolation of specific multimeric protein assemblies, allowing purification of specific complexes according to both the number and the position of charged peptide tags.

Gibbs-Donnan Effect

In ion exchange chromatography, the Gibbs–Donnan effect is observed when the pH of the applied buffer and the ion exchanger differ, even up to one pH unit. For example, in anion-exchange columns, the ion exchangers repeal protons so the pH of the buffer near the column differs is higher than the rest of the solvent. As a result, an experimenter has to be careful that the protein(s) of interest is stable and properly charged in the "actual" pH.

This effect comes as a result of two similarly charged particles, one from the resin and one from the solution, failing to distribute properly between the two sides; there is a selective uptake of one ion over another. For example, in a sulphonated polystyrene resin, a cation exchange resin, the chlorine ion of a hydrochloric acid buffer should equilibrate into the resin. However, since the concentration of the sulphonic acid in the resin is high, the hydrogen of HCl has no tendency to enter the column. This, combined with the need of electroneutrality, leads to a minimum amount of hydrogen and chlorine entering the resin.

Uses

Clinical Utility

A use of ion chromatography can be seen in the argentation ion chromatography. Usually silver and compounds containing acetylenic and ethylenic bonds have very weak interactions. This phenomenon has been widely tested on olefin compounds. The ion complexes the olefins make with silver ions are weak and made based on the overlapping of pi, sigma, and d orbitals and available electrons therefore cause no real changes in the double bond. This behavior was manipulated to separate lipids, mainly fatty acids from mixtures in to fractions with differing number of double bonds using silver ions. The ion resins were impregnated with silver ions, which were then exposed to various acids (silicic acid) to elute fatty acids of different characteristics.

Detection limits as low as 1 µM can be obtained for alkali metal ions. It may be used for measurement of HbA1c, porphyrin and with water purification. Ion Exchange Resins(IER) have been widely used especially in medicines due to its high capacity and the uncomplicated system of the separation process. One of the synthetic uses is to use Ion Exchange Resins for kidney dialysis. This method is used to separate the blood elements by using the cellulose membraned artificial kidney.

Another clinical application of ion chromatography is in the rapid anion exchange chromatography technique used to separate creatine kinase (CK) isoenzymes from human serum and tissue sourced in autopsy material (mostly CK rich tissues were used such as cardiac muscle and brain). These isoenzymes include MM, MB, and BB, which all carry out the same function given different amino acid sequences. The functions of these isoenzymes are to convert creatine, using ATP, into phosphocreatine expelling ADP. Mini columns were filled with DEAE-Sephadex A-50 and further eluted with tris- buffer sodium chloride at various concentrations (each concentration was chosen advantageously to manipulate elution). Human tissue extract was inserted in columns for separation. All fractions were analyzed to see total CK activity and it was found that each source of CK isoenzymes had characteristic isoenzymes found within. Firstly, CK-MM was eluted, then CK-MB, followed by CK-BB. Therefore, the isoenzymes found in each sample could be used to identify the source, as they were tissue specific.

Using the information from results, correlation could be made about the diagnosis of patients and the kind of CK isoenzymes found in most abundant activity. From the finding, about 35 out of 71 patients studied suffered from heart attack (myocardial infarction) also contained an abundant amount of the CK-MM and CK-MB isoenzymes. Findings further show that many other diagnosis including renal failure, cerebrovascular disease, and pulmonary disease were only found to have the CK-MM isoenzyme and no other isoenzyme. The results from this study indicate correlations between various diseases and the CK isoenzymes found which confirms previous test results using various techniques. Studies about CK-MB found in heart attack victims have expanded since this study and application of ion chromatography.

Industrial Applications

Since 1975 ion chromatography has been widely used in many branches of industry. The main beneficial advantages are reliability, very good accuracy and precision, high selectivity, high speed, high separation efficiency, and low cost of consumables. The most significant development related to ion chromatography are new sample preparation methods; improving the speed and selectivity of analytes separation; lowering of limits of detection and limits of quantification; extending the scope of applications; development of new standard methods; miniaturization and extending the scope of the analysis of a new group of substances. Allows for quantitative testing of electrolyte and proprietary additives of electroplating baths. It is an advancement of qualitative hull cell testing or less accurate UV testing. Ions, catalysts, brighteners and accelerators can be measured. Ion exchange chromatography has gradually become a widely known, universal technique for the detection of both anionic and cationic species. Applications for such purposes have been developed, or are under development, for a variety of fields of interest, and in particular, the pharmaceutical industry. The usage of ion exchange chromatography in pharmaceuticals has increased in recent years, and in 2006, a chapter on ion exchange chromatography was officially added to the United States Pharmacopia-National Formulary (USP-NF). Furthermore, in 2009 release of the USP-NF, the United States Pharmacopia made several analyses of ion chromatography available using two techniques: conductivity detection, as well as pulse amperometric detection. Majority of these applications are primarily used for measuring and analyzing residual limits in pharmaceuticals, including detecting the limits of oxalate, iodide, sulfate, sulfamate, phosphate, as well as various electrolytes including potassium, and sodium. In total, the 2009 edition of the USP-NF officially released twenty eight methods of detection for the analysis of active compounds, or components of active compounds, using either conductivity detection or pulse amperometric detection.

Drug Development

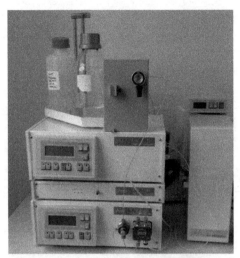

An ion chromatography system used to detect and measure cations such as sodium, ammonium and potassium in Expectorant Cough Formulations

There has been a growing interest in the application of IC in the analysis of pharmaceutical drugs. IC is used in different aspects of product development and quality control testing. For example, IC is used to improve stabilities and solubility properties of pharmaceutical active drugs molecules as well as used to detect systems that have higher tolerance for organic solvents. IC has been used for the determination of analytes as a part of a dissolution test. For instance, calcium dissolution tests have shown that other ions present in the medium can be well resolved among themselves and also from the calcium ion. Therefore, IC has been employed in drugs in the form of tablets and capsules in order to determine the amount of drug dissolve with time. IC is also widely used for detection and quantification of excipients or inactive ingredients used in pharmaceutical formulations. Detection of sugar and sugar alcohol in such formulations through IC has been done due to these polar groups getting resolved in ion column. IC methodology also established in analysis of impurities in drug substances and products. Impurities or any components that are not part of the drug chemical entity are evaluated and they give insights about the maximum and minimum amounts of drug that should be administered in a patient per day.

Ion Exchange Chromatograph

Ion exchange chromatography is a fast, economical and versatile technique for effective separation of ions, amino acids, peptides, nucleotide and nucleic acids etc. This technique is widely used in the pre-fractionation or purification of a target protein from crude biological samples. Before we go in details of ion-exchange chromatography of proteins, let us discuss, ionization and charge on proteins with respect to pH.

> **Concept of charge on Protein:** A typical protein contains several ionizable group (basic amino acid side chains of lysine, arginine and histidine as well as acidic side chains of glutamate and aspartate). Additionally, N-terminal amino group and C-terminal carboxy group can also ionize. Ionization state of these amino acid sside chains depends on pK (also depainds on localized environ ment of side chain) and pH which is described by the Henderson-hasselbalch equation.
>
> $$pH = pK + \log\frac{[\text{conjugate base}]}{[\text{conjugate acid}]}$$
>
> Let us try to understand effect of Lysine on net charge on protein. Charge due to Lysine can range between 0 (when pH >> pK_A) and +1 (when pH << pK_A). The side chain on lysine normally has a pK value of approx 9.0 (the precise value depends on side chain environment). When pH = pK, the concentration of the protonated and deprotonated forms is equal and the charge due to Lysine is +0.5. When pH = 11.0, the lysine is about 99% deprotonated, leading to a charge of +0.01. When pH = 7.0, the lysine is about 99% protonated, leading to a charge of +0.99.
>
> At a given pH over all charge on protein depends on sum of charges on indivisual amino acid side chance as well as C- and N-terminal. pH value with net zero charge is called isolelectric point. As proteins in a crude mixture vary in terms of sequence and amino acid compositions, they are likely to have different net charge at a given pH value.

Principle of Ion Exchange Chromatography: Ion exchange chromatography separates proteins or other molecules based on differences in their accessible surface charges. In ion exchange chromatography the analyte molecules are retained on the column based on coulombic (ionic) interactions. The stationary phase surface contains ionic functional groups of opposite charge that interact with analyte ions. The elution is done by increasing salt gradient. Most commonly used salt is NaCl, exists in equilibrium with Na^+ (cation) and Cl^- (anion) in aqueous solution. As the concentration of salt increases concentration of Na^+ (cation) and Cl^- (anion) also increases. The basic principle of ion exchange chromatography is the reversible exchange of analyte ions bound to solid support with similar ions generated from salt in liquid phase. Many biological molecules such as proteins, amino acids, nucleotides and other ions have ionisable groups which carries a net charge (positive or negative) dependent on their pKa and on the pH of the solution, which can be utilised in separating mixture of such molecules as explained in box. Ion exchange chromatography experiments are carried out mainly in columns packed with ion exchangers. On the basis of type of exchanger used for separation this chromatography is further subdivided into cation exchange chromatography and anion exchange chromatography.

Many biological molecules, especially proteins, are stable within a narrow pH range so the type of exchanger selected must operate within this range. Suppose if protein is most stable below its isolecteic point (pI), there will be net positive charge on the protein surface, so for separation of this protein cation exchanger should be used (experimental pH value should be between lowest pH wehere protein is stable and pI value). If protein is most stable above its pI, there will be net negative charge on the protein surface and anion exchanger should be used (experimental pH value should be between highest pH where protein is stable and pI value). If protein is stable over a wide range of pH, it can be separated by either type of ion exchanger (experimental pH value may be decided considering lowest and highest pH value stability of the protein). Weak electrolyte requires very high or very low pH for ionisation so it can only be separated on strong exchanger, as they only operate over a wide pH range, whereas in case of strong electrolytes, weak exchangers are preferred.

Anion Exchange Chromatography: Anion exchange chromatography exploits difference in surface negative charge of protein or other molecules for separation. Anion exchange matrix has positively charged functional group. Few common examples of anion exchanger functional groups are as follows

Diethylaminoethyl (DEAE)　　　　　$\text{—O–CH}_2\text{–CH}_2\text{–N}^{\pm}\text{H}$ with $\text{CH}_2\text{–CH}_3$ groups

Quaternary aminoethyl (QAE)　　$-\text{O–CH}_2\text{–CH}_2\text{–N}^+(\text{C}_2\text{H}_5)_2\text{–CH}_2\text{–CHOH–CH}_3$

Quaternary ammonium (Q)　　　　$\text{—O–N}^{\oplus}\text{R}_3$

As explained in cation exchange chromatography, solid support with these functional groups can be prepared with various beads. They differ in few properties like flow rate etc. Anion exchangers based on dextran (Sephadex), agarose (Sepharose) or cross-linked cellulose (Sephacel) are few common options. Anion exchange Chromatography are performed using buffers at pH between 7 and 10 and running a gradient from a solution containing just this buffer to a solution containing this buffer with 1M NaCl. The surface charge of the molecule (proteins, nucleic acids etc) which bind will be net negative, thus to get binding of a specific protein one should perform purification above the pI of that protein.

Q- Anion Exchanger　　　　DEAE Anion Exchanger

The salt in the solution competes for the binding to the column matrix and releases the protein from its bound state at a given concentration. Proteins separate because the amount of salt needed to release the protein varies with the external charge of the protein.

Size-exclusion Chromatography

Size-exclusion chromatography (SEC), also known as molecular sieve chromatography, is a chromatographic method in which molecules in solution are separated by their size, and in some cases molecular weight. It is usually applied to large molecules or macromolecular complexes such as proteins and industrial polymers. Typically, when an aqueous solution is used to transport the sample through the column, the technique is known as gel-filtration chromatography, versus the name gel permeation chromatography, which is used when an organic solvent is used as a mobile phase. SEC is a widely used polymer characterization method because of its ability to provide good molar mass distribution (Mw) results for polymers.

Applications

The main application of gel-filtration chromatography is the fractionation of proteins and other water-soluble polymers, while gel permeation chromatography is used to analyze the molecular weight distribution of organic-soluble polymers. Either technique should not be confused with gel electrophoresis, where an electric field is used to "pull" or "push" molecules through the gel depending on their electrical charges.

Advantages

The advantages of this method include good separation of large molecules from the small molecules with a minimal volume of eluate, and that various solutions can be applied without interfering with the filtration process, all while preserving the biological activity of the particles to separate. The technique is generally combined with others that further separate molecules by other characteristics, such as acidity, basicity, charge, and affinity for certain compounds. With size exclusion chromatography, there are short and well-defined separation times and narrow bands, which lead to good sensitivity. There is also no sample loss because solutes do not interact with the stationary phase.

The other advantage to this experimental method is that in certain cases, it is feasible to determine the approximate molecular weight of a compound. The shape and size of the compound (eluent) determine how the compound interacts with the gel (stationary phase). To determine approximate molecular weight, the elution volumes of compounds with their corresponding molecular weights are obtained and then a plot of "K_{av}" vs "log(Mw)" is made, where $K_{av} = (V_e - V_o)/(V_t - V_o)$ and Mw is the molecular mass. This plot acts as a calibration curve, which is used to approximate the desired compound's molecular weight. The V_e component represents the volume at which the intermediate molecules elute such as molecules that have partial access to the beads of the column. In addition, V_t is the sum of the total volume between the beads and the volume within the beads. The V_o component represents the volume at which the larger molecules elute, which elute in the beginning. Disadvantages are, for example, that only a limited number of bands can be accommodated because the time scale of the chromatogram is short, and, in general, there must be a 10% difference in molecular mass to have a good resolution.

Discovery

The technique was invented by Grant Henry Lathe and Colin R Ruthven, working at Queen Charlotte's Hospital, London. They later received the John Scott Award for this invention. While Lathe and Ruthven used starch gels as the matrix, Jerker Porath and Per Flodin later introduced dextran gels; other gels with size fractionation properties include agarose and polyacrylamide. A short review of these developments has appeared.

There were also attempts to fractionate synthetic high polymers; however, it was not until 1964, when J. C. Moore of the Dow Chemical Company published his work on the preparation of gel permeation chromatography (GPC) columns based on cross-linked polystyrene with controlled pore size, that a rapid increase of research activity in this field began. It was recognized almost immediately that with proper calibration, GPC was capable to provide molar mass and molar mass distribution information for synthetic polymers. Because the latter information was difficult to obtain by other methods, GPC came rapidly into extensive use.

Theory and Method

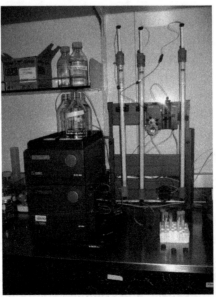

Agarose-based SEC columns used for protein purification on an AKTA FPLC machine

SEC is used primarily for the analysis of large molecules such as proteins or polymers. SEC works by trapping smaller molecules in the pores of the adsorbent materials adsorption ("stationary phases"). This process is usually performed with a column, which consists of a hollow tube tightly packed with extremely small porous polymer beads designed to have pores of different sizes. These pores may be depressions on the surface or channels through the bead. As the solution travels down the column some particles enter into the pores. Larger particles cannot enter into as many pores. The larger the particles, the faster the elution. The larger molecules simply pass by the pores because those molecules are too large to enter the pores. Larger molecules therefore flow through the column more quickly than smaller molecules, that is, the smaller the molecule, the longer the retention time.

One requirement for SEC is that the analyte does not interact with the surface of the stationary phases, with differences in elution time between analytes ideally being based solely on the solute volume the analytes can enter, rather than chemical or electrostatic interactions with the stationary phases. Thus, a small molecule that can penetrate every region of the stationary phase pore system can enter a total volume equal to the sum of the entire pore volume and the interparticle volume. This small molecule elutes late (after the molecule has penetrated all of the pore- and interparticle volume—approximately 80% of the column volume). At the other extreme, a very large molecule that cannot penetrate any the smaller pores can enter only the interparticle volume (~35% of the column volume) and elutes earlier when this volume of mobile phase has passed through the column. The underlying principle of SEC is that particles of different sizes elute (filter) through a stationary phase at different rates. This results in the separation of a solution of particles based on size. Provided that all the particles are loaded simultaneously or near-simultaneously, particles of the same size should elute together.

However, as there are various measures of the size of a macromolecule (for instance, the radius of gyration and the hydrodynamic radius), a fundamental problem in the theory of SEC has been the choice of a proper molecular size parameter by which molecules of different kinds are separated. Experimentally, Benoit and co-workers found an excellent correlation between elution volume and a dynamically based molecular size, the hydrodynamic volume, for several different chain architecture and chemical compositions. The observed correlation based on the hydrodynamic volume became accepted as the basis of universal SEC calibration.

Still, the use of the hydrodynamic volume, a size based on dynamical properties, in the interpretation of SEC data is not fully understood. This is because SEC is typically run under low flow rate conditions where hydrodynamic factor should have little effect on the separation. In fact, both theory and computer simulations assume a thermodynamic separation principle: the separation process is determined by the equilibrium distribution (partitioning) of solute macromolecules between two phases --- a dilute bulk solution phase located at the interstitial space and confined solution phases within the pores of column packing material. Based on this theory, it has been shown that the relevant size parameter to the partitioning of polymers in pores is the mean span dimension (mean maximal projection onto a line). Although this issue has not been fully resolved, it is likely that the mean span dimension and the hydrodynamic volume are strongly correlated.

A size exclusion column

Each size exclusion column has a range of molecular weights that can be separated. The exclusion limit defines the molecular weight at the upper end of the column 'working' range and is where molecules are too large to get trapped in the stationary phase. The lower end of the range is defined by the permeation limit, which defines the molecular weight of a molecule that is small enough to penetrate all pores of the stationary phase. All molecules below this molecular mass are so small that they elute as a single band.

The filtered solution that is collected at the end is known as the eluate. The void volume includes any particles too large to enter the medium, and the solvent volume is known as the column volume.

Factors Affecting Filtration

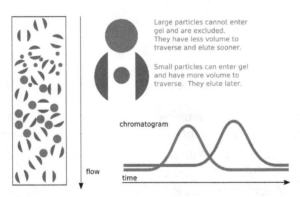

A cartoon illustrating the theory behind size exclusion chromatography

In real-life situations, particles in solution do not have a fixed size, resulting in the probability that a particle that would otherwise be hampered by a pore passing right by it. Also, the stationary-phase particles are not ideally defined; both particles and pores may vary in size. Elution curves, therefore, resemble Gaussian distributions. The stationary phase may also interact in undesirable ways with a particle and influence retention times, though great care is taken by column manufacturers to use stationary phases that are inert and minimize this issue.

Like other forms of chromatography, increasing the column length enhances resolution, and increasing the column diameter increases column capacity. Proper column packing is important for maximum resolution: An over-packed column can collapse the pores in the beads, resulting in a loss of resolution. An under-packed column can reduce the relative surface area of the stationary phase accessible to smaller species, resulting in those species spending less time trapped in pores. Unlike affinity chromatography techniques, a solvent head at the top of the column can drastically diminish resolution as the sample diffuses prior to loading, broadening the downstream elution.

Analysis

In simple manual columns, the eluent is collected in constant volumes, known as fractions. The more similar the particles are in size the more likely they are in the same fraction and not detected separately. More advanced columns overcome this problem by constantly monitoring the eluent.

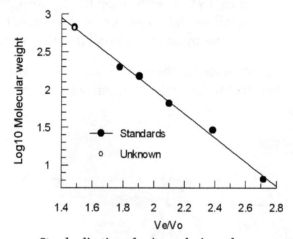

Standardization of a size exclusion column

The collected fractions are often examined by spectroscopic techniques to determine the concentration of the particles eluted. Common spectroscopy detection techniques are refractive index (RI) and ultraviolet (UV). When eluting spectroscopically similar species (such as during biological purification), other techniques may be necessary to identify the contents of each fraction. It is also possible to analyse the eluent flow continuously with RI, LALLS, Multi-Angle Laser Light Scattering MALS, UV, and/or viscosity measurements.

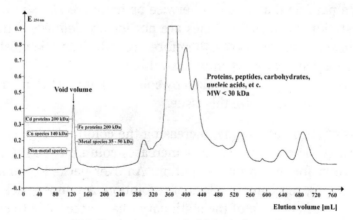

SEC Chromatogram of a biological sample

The elution volume (Ve) decreases roughly linear with the logarithm of the molecular hydrodynamic volume. Columns are often calibrated using 4-5 standard samples (e.g., folded proteins of known molecular weight), and a sample containing a very large molecule such as thyroglobulin to determine the void volume. (Blue dextran is not recommended for Vo determination because it is heterogeneous and may give variable results) The elution volumes of the standards are divided by the elution volume of the thyroglobulin (Ve/Vo) and plotted against the log of the standards' molecular weights.

Applications

Biochemical Applications

In general, SEC is considered a low resolution chromatography as it does not discern similar species very well, and is therefore often reserved for the final "polishing" step of a purification. The technique can determine the quaternary structure of purified proteins that have slow exchange times, since it can be carried out under native solution conditions, preserving macromolecular interactions. SEC can also assay protein tertiary structure, as it measures the hydrodynamic volume (not molecular weight), allowing folded and unfolded versions of the same protein to be distinguished. For example, the apparent hydrodynamic radius of a typical protein domain might be 14 Å and 36 Å for the folded and unfolded forms, respectively. SEC allows the separation of these two forms, as the folded form elutes much later due to its smaller size.

Polymer Synthesis

SEC can be used as a measure of both the size and the polydispersity of a synthesised polymer, that is, the ability to find the distribution of the sizes of polymer molecules. If standards of a known size are run previously, then a calibration curve can be created to determine the sizes of polymer molecules of interest in the solvent chosen for analysis (often THF). In alternative fashion, techniques such as light scattering and/or viscometry can be used online with SEC to yield absolute molecular weights that do not rely on calibration with standards of known molecular weight. Due to the difference in size of two polymers with identical molecular weights, the absolute determination methods are, in general, more desirable. A typical SEC system can quickly (in about half an hour) give polymer chemists information on the size and polydispersity of the sample. The preparative SEC can be used for polymer fractionation on an analytical scale.

Drawback

In SEC, mass is not measured so much as the hydrodynamic volume of the polymer molecules, that is, how much space a particular polymer molecule takes up when it is in solution. However, the approximate molecular weight can be calculated from SEC data because the exact relationship between molecular weight and hydrodynamic volume for polystyrene can be found. For this, polystyrene is used as a standard. But the relationship between hydrodynamic volume and molecular weight is not the same for all polymers, so only an approximate measurement can be obtained. Another drawback is the possibility of interaction between the stationary phase and the analyte. Any interaction leads to a later elution time and thus mimics a smaller analyte size.

When performing this method, the bands of the eluting molecules may be broadened. This can occur by turbulence caused by the flow of the mobile phase molecules passing through the molecules of the stationary phase. In addition, molecular thermal diffusion and friction between the molecules of the glass walls and the molecules of the eluent contribute to the broadening of the bands. Besides broadening, the bands also overlap with each other. As a result, the eluent usually gets considerably diluted. A few precautions can be taken to prevent the likelihood of the bands broadening. For instance, one can apply the sample in a narrow, highly concentrated band on the top of the column. The more concentrated the eluent is, the more efficient the procedure would be. However, it is not always possible to concentrate the eluent, which can be considered as one more disadvantage.

Absolute size-exclusion Chromatography

Absolute size-exclusion chromatography (ASEC) is a technique that couples a dynamic light scattering (DLS) instrument to a size exclusion chromatography system for absolute size measurements of proteins and macromolecules as they elute from the chromatography system.

The definition of absolute used here is that it does not require calibration to obtain hydrodynamic size, often referred to as hydrodynamic diameter (D_H in units of nm). The sizes of the macromolecules are measured as they elute into the flow cell of the DLS instrument from the size exclusion column set. It should be noted that the hydrodynamic size of the molecules or particles are measured and not their molecular weights. For proteins a Mark-Houwink type of calculation can be used to estimate the molecular weight from the hydrodynamic size.

A big advantage of DLS coupled with SEC is the ability to obtain enhanced DLS resolution. Batch DLS is quick and simple and provides a direct measure of the average size, but the baseline resolution of DLS is 3 to 1 in diameter. Using SEC, the proteins and protein oligomers are separated, allowing oligomeric resolution. Aggregation studies can also be done using ASEC. Though the aggregate concentration may not be calculated, the size of the aggregate can be measured, only limited by the maximum size eluting from the SEC columns.

Limitations of ASEC include flow-rate, concentration, and precision. Because a correlation function requires anywhere from 3–7 seconds to properly build, a limited number of data points can be collected across the peak.

References

- Hanaor, D.A.H.; Michelazzi, M.; Leonelli, C.; Sorrell, C.C. (2012). "The effects of carboxylic acids on the aqueous dispersion and electrophoretic deposition of ZrO_2". Journal of the European Ceramic Society. 32 (1): 235–244. doi:10.1016/j.jeurceramsoc.2011.08.015

- Kindt, Thomas; Goldsby, Richard; Osborne, Barbara (2007). Kuby Immunology. New York: W.H. Freeman and Company. p. 553. ISBN 978-1-4292-0211-4

- Song, Di; Ma, Shang; Khor, Soo Peang (2002-01-01). "Gel electrophoresis-autoradiographic image analysis of radiolabeled protein drug concentration in serum for pharmacokinetic studies". Journal of Pharmacological and Toxicological Methods. 47 (1): 59–66. ISSN 1056-8719. PMID 12387940

- J., Ninfa, Alexander; P., Ballou, David (2004). Fundamental laboratory approaches for biochemistry and biotechnology. Wiley. ISBN 9781891786006. OCLC 633862582

- Booth, F. (1948). "Theory of Electrokinetic Effects". Nature. 161 (4081): 83–86. Bibcode:1948Natur.161...83B. PMID 18898334. doi:10.1038/161083a0

- Minde DP (2012). "Determining biophysical protein stability in lysates by a fast proteolysis assay, FASTpp". PLOS ONE. 7 (10): e46147. PMC 3463568 . PMID 23056252. doi:10.1371/journal.pone.0046147

- Mikkelsen, Susan; Cortón, Eduardo (2004). Bioanalytical Chemistry. John Wiley & Sons, Inc. p. 224. ISBN 0-471-62386-5

- O'Brien, R.W.; L.R. White (1978). "Electrophoretic mobility of a spherical colloidal particle". J. Chem. Soc. Faraday Trans. 2 (74): 1607. doi:10.1039/F29787401607

- Fritz, J. S. (2004). "Early milestones in the development of ion-exchange chromatography: a personal account". Journal of chromatography. A. 1039 (1–2): 3–12. PMID 15250395. doi:10.1016/s0021-9673(04)00020-2

- Ninfa, Alexander; Ballou, David; Benore, Marilee (May 26, 2009). Fundamental Laboratory Approaches for Biochemistry and Biotechnology. Wiley. pp. 143–145. ISBN 0470087668

- Lucy, C. A. (2003). "Evolution of ion-exchange: from Moses to the Manhattan Project to Modern Times". Journal of chromatography. A. 1000 (1–2): 711–24. PMID 12877196. doi:10.1016/s0021-9673(03)00528-4

- Appling, Dean; Anthony-Cahill, Spencer; Mathews, Christopher (2016). Biochemistry: Concepts and Connections. New Jersey: Pearson. p. 134. ISBN 9780321839923

- Laemmli UK (August 1970). "Cleavage of structural proteins during the assembly of the head of bacteriophage T4". Nature. 227 (5259): 680–685. PMID 5432063. doi:10.1038/227680a0

- Kirkland, J. J.; DeStefano, J. J. (2006-09-08). "The art and science of forming packed analytical high-performance liquid chromatography columns". Journal of Chromatography A. The Role of Theory in Chromatography. 1126 (1–2): 50–57. doi:10.1016/j.chroma.2006.04.027

- Bhattacharyya, Lokesh; Rohrer, Jeffrey (2012). Applications of Ion Chromatography in the Analysis of Pharmaceutical and Biological Products. Wiley. p. 247. ISBN 0470467096

An Overview of Mass Spectrometry

Mass spectrometry evaluates the mass-to-charge ratio of a particle existing in vacuum. Upon finding the mass-to-charge ratio, particles can be separated, which helps in conducting many experiments. The aspects elucidated in this chapter are of vital importance, and provide a better understanding of mass spectrometry.

Mass Spectrometry

Mass spectrometry (MS) is an analytical technique that measures the mass/charge ratio of charged particles in vacuum. Mass spectrometry can determine masse/charge ratio with high accuracy. Molecules in a test sample are converted to gaseous ions that are subsequently separated according to mass/charge ratio. Several types of experiments can be performed with mass spectrometry.

A typical mass spectrometry instrument has three components as shown in Figure.

1. Ion source
2. Analyzer
3. Detector: The detector records the current produced when an ion passes by or hits a surface. Several types of detector are used like electron multiplier, Faraday cups and ion to photon detectors.

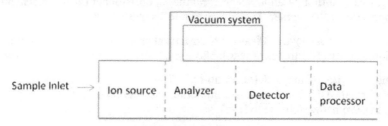

Basic components of mass spectrometry

All mass spectrometry operate under vacuum (10^{-6} torr pressure). Without high vacuum, the ions produced in the source will not reach the detector. At atmospheric pressure, the mean free path of a typical ion is around 52 nm; at 1mtorr, it is 40 mm; and at 10^{-6} torr, it is 40 m. Sample inlet may be coupled to a liquid chromatography system (call LC-MS), gas chromatography system (GC-MS) or capillary electrophoresis system.

An Overview of Mass Spectrometry

We shall study components of mass spectrometry (MS) in detail. A simplified scheme of mass spectrometry (MS) and MS-MS (tandem mass spectrometry) is shown in Figure.

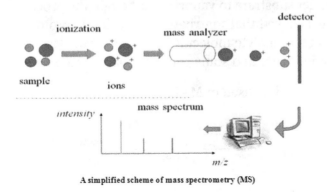

A simplified scheme of mass spectrometry (MS)

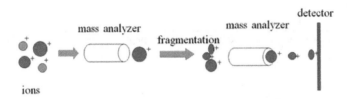

A simplified scheme of MS-MS

A simplified scheme of mass spectrometry and Tandem mass spectrometry (also called MS MS). In MS-MS, after first analyzer, analytes are fragmented and fragments are analyzed in second analyzer.

Ion source:

There are several types of ionization methods in mass spectrometry. The physical basis of ionization methods are very complex and outside the scope of the course. Most common methods are:

(a) Matrix-assisted laser desorption/ionization (MALDI)

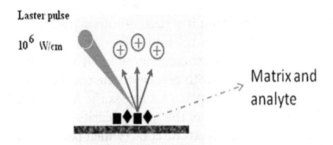

Scheme of Matrix-assisted laser desorption/ionization (MALDI)

This method of ionization is a soft ionization method and results in minimum fragmentation of sample. This method is used for non-volatile, and thermally labile compounds such as proteins, oligonucleotides, synthetic polymers. Sample is mixed with

1000 times molar excess of sample and spotted onto a metal plate and dried. Matrix plays a key role in this technique by absorbing the laser light energy and causing a small part of the target substrate to vaporize. Although, the process of forming analyte ions is unclear, it is believed that matrix which has labile protons, such as carboxylic acids, protonates neutral analyte molecules after absorbing laser light energy. Scheme of MALDI is shown in Figure and some common matrices are listed in Table.

Table: Few common matrices used in MALDI

Few common matrices used in MALDI

Matrix	Application
3,5-Dimethoxy-4-hydroxycinnamic acid 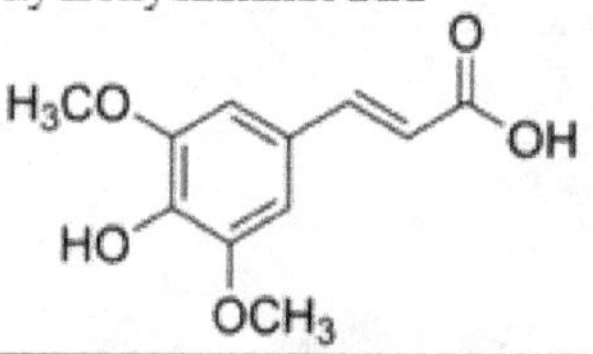	Higher mass biopolymer
α-cyano-4-hydroxycinnamic acid 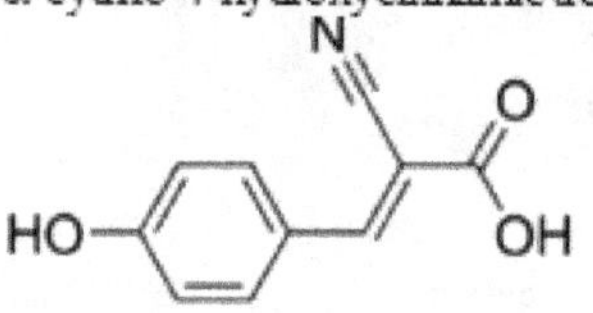	Protein, peptides, organic compounds
2,5 dihydroxy benzoic acid	Oligonucleotide
Trihydroxyacetophenone	Peptides, Oligonucleotide

Electrospray Ionisation (ESI)

Electrospray Ionisation (ESI) is a preferred method of ionization when the sample is in liquid form. This is also a soft method of ionization and results in less fragmentation. ESI is a very valuable method for analysis of biological samples. The method was developed by John Fenn and he shared 2002 Nobel prize in chemistry for this work. The analyte is introduced either from a syringe pump or as the eluent flow from liquid chromatography with a flow rate 1μl min^{-1}. The analyte solution passes through the electrospray needle (Stainless steel capillary with 75-150 1μm internal diameters) that has a high potential difference (with respect to the counter electrode) applied to it (typically in the range from 2.5 to 4 kV). This forces the spraying of charged droplets from the needle with a surface charge of the same polarity to the charge on the needle. As droplet moves towards counter electrode cone (which passes it to analyzer), solvent evaporation occurs and droplet shrinks until it reaches the point that the surface tension can no longer sustain the charge (the Rayleigh limit) and at that point droplets break. This produces smaller droplets and the process is repeated. Finally after all solvent evaporated, charge is passed on to analyte. These charged analyte molecules can have single or multiple charges.

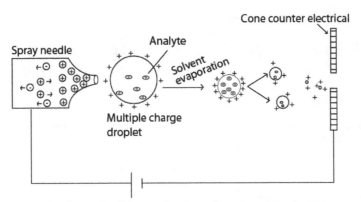

A schematic of the mechanism of ion formation in ESI.

Electron Ionization

Electron Ionization (EI) works well for many gas phase molecules, but it results in extensive fragmentation and molecular ions are not observed for many compounds. Fragmentation mass spectra are sometime useful because it provides structural information of a molecule.

The electron beam is produced by a filament of rhenium or tungsten wire by thermionic emission. When cathode filament of rhenium or tungsten is heated at temperature over 1000 K, electrons are emitted. The generated electrons are accelerated to 70 eV which results in electron beam. The volatile sample or sample in gaseous phase containing neutral molecules is introduced to the ion source in a perpendicular direction to the electron beam. Electron impact on the analyte results in either loss of electron (to produce cation) or gain of electron (to produce anion). Chemical bonds in organic molecules are formed by pairing of electrons. Electron impact may knock out one of the electron. This leaves the bond with a single unpaired electron. This is radical as well as being cation written as $M^{+\cdot}$, where (+) indicates ionic state while (.) indicates radical. Electron impact may result in electron capture (extra unpaired electron). This generates a radical as well as being anion written as $M^{+\cdot}$, where (-) indicates ionic state while (.) indicates radical.

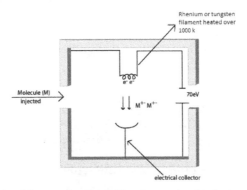

A schematic of the mechanism of ion formation in electron ionization

Chemical Ionization

Set-up for chemical Ionization is similar to electron impact ionization. However, in this method, a reagent gas like CH4 is injected in the ion chamber. Due to electron impact, the reagent gas in the chemical ionization source gets ionized. This follows injection of analyte molecule. Analyte molecules undergo many collisions with the reagent gas. The reagent gas ions in this cloud react and produce adduct ions as shown in figure, which are excellent proton donors for analyte.

Electron impact on reagent gass

$$CH_4 + e^- \longrightarrow CH_4^+ + 2e^-$$

Reaction of reagent gas to form 1 on

$$CH_4^+ + CH_4 \longrightarrow CH_3 + CH^+5$$

Reaction of Reagent gas 1 on with analyte

$$CH_5^+ = M \longrightarrow CH_4 + CH_4 + MH^+$$

Most common types of analyzers are :

1. Time of Flight (ToF)
2. Magnetic sector
3. Quadrapole

Time of Flight (ToF) Analyzer

Once ions are generated in the ion source, they are extracted through extraction grid and accelerated to ToF analyzer by acceleration grid (has high negative potential) as shown in Figure.

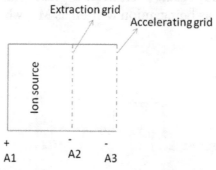

A2 is 100 volts more negative then A1; A3 is 1000 volts more negative then A2
Time of Flight (ToF) analyzer showing extraction and accelerating grids

If an ion with charge q (please note: some time charge is written as z) is accelerated with potential V

$$qV = 1/2mv^2$$

$$v2 = 2qV/m$$

If the flight tube length is L, time taken to reach at the end of tube (detector) is

$$t = \left(\frac{m}{2qV_0}\right)^{1/2} L$$

Where

t = time – of – flight (s)

m = mass of the ion (kg)

q = charge on ion (C)

V_0 = accelerating potential (V)

If L is not large, time difference for close mass/charge ions will be less and spectra will overlap. This results in low resolution.

Thus, for a better resolution, L should be large. At the same time there is a demand for miniaturized instrument. Thus, ion reflectors are used to increase flight tube length. A simple setup for Time of Flight (ToF) analyzer is shown in Figure and Time of Flight (ToF) analyzer with ion reflectors is shown in Figure.

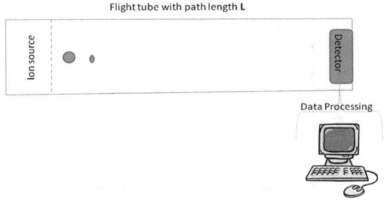

A simple setup for Time of Flight (ToF) analyze

As shown in the Figure, ion reflector has a decelerating electrode and reflecting electrode. In a typical mass spectrometric instrument flight tube is small and does not give good resolution as the time difference between different ions are less. A linear flight tube (also called linear mode) is good when a single ion is being analyzed or mass/charge difference between ions being analyzed are not very close. Thus, there is a requirement of better resolution to analyze close mass/charge ions. This may be achieved by increasing path length with the help of an ion reflector (reflectron mode). An advantage to the reflectron mode −ToF of arrangement is that larger flight path is achieved in a given length of instrument improving overall resolution.

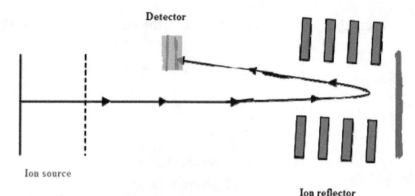

A simple setup for Time of Flight (ToF) analyzer with ion reflectors

As soon as an ion reaches the detector, time taken to reach the detector is noted. As light path of an instrument (L) and accelerating potential (V) of an instrument is known, software will calculate mass/charge ration and plot a signal at respective mass and charge. As mass spectrometry is only a qualitative method (not a quantitative method) intensity value in mass spectrometry data can't be used for calculation of relative abundance of ions with different mass/charge ratio.

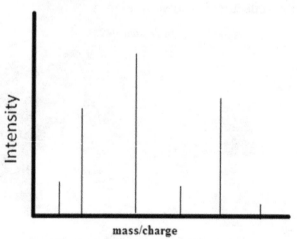

A pictorial representation of mass spectrometry data

Let us take an example of a peptide of mass 500 Da ionized by MALDI (protonation).

It may give a signal at Peptide + one proton mass/one charge =501/1=501.

In case there are two proton, the same peptide may give a signal at Peptide + two proton mass/two charge =502/2=251.

Thus a homogenous peptide may give multiple peaks.

Magnetic Sector Analyzer

Ion with charge z traveling with velocity v in a magnetic field with magnitude B (perpendicular to travel path) experiences a torque = Bzv.

Mass of the particle has no effect on torque and there will not be any if the particle is stationary. The result of the torque is that the path of the ion will become curved. Incase, magnetic field is perpendicular to path of ion at all the points, ion will have a circular path. To find the radius of curvature we can equate the

$$\text{Centrifugal force} = mv^2/r$$

$$Bzv = \frac{mv^2}{r}$$

$$r = \frac{mv^2}{Bzv} = \frac{v\,m}{B\,z}$$

The velocity, v, of the ion is usually the result of the acceleration of the ion through an electric field, so that the v is given by:

$$\frac{1}{2}mv2 = zV$$

$$v = \sqrt{2v\frac{z}{m}}$$

where V is the electric potential that causes acceleration. By substituting this expression in the one for r, we obtain

$$r = \frac{v\,m}{B\,z} = \frac{\sqrt{2v\frac{z}{m}}\,m}{B\quad z} = \frac{\sqrt{2v}}{B}\sqrt{\frac{m}{z}}$$

As clear from the equation, mass to charge ratio decides the path radius of ion in the magnetic field. Particles with the same mass to charge ratio will follow the same path in the magnetic field.

From above discussion it is clear that at a given magnetic field strength a particular mass/charge ion can reach the detector. When an unknown sample is analyzed, magnetic field is varied to get signal at detector. Once we know the magnetic field strength giving signal, mass/charge can be calculated.

In case of magnetic sector analyzer, magnetic field strength (B) is varied while accelerating potential (V) and radius (r) of the analyzer is known and remains constant during experiment. As soon as an ion reaches the detector, magnetic field strength (B) at that time is noted. As accelerating potential (V) and radius (r) of the analyzer is known and remains constant during experiment, software will calculate mass/charge ratio that can reach the detector and plot a signal at respective mass and charge.

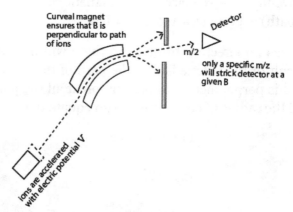

A simple setup for Magnetic sector analyzer

Quadrapole

Quadrapole analyzer was discovered by Wolfgang Paul who shared Nobel prize of Physics in 1989 for this work with Hans Georg Dehmelt. As the name implies, it consists of 4 straight and parallel rods, set parallel to each other as shown in the figure A voltage of a DC component (U) and a radio frequency component (Vcos ωt) is applied between adjacent rods, opposite rods being electrically connected. Ions oscillate inside quadruple in the (x), and (y) dimensions as a result of the high frequency electric field.

Two parameters, (a) and (q) determines the stability of the oscillating ions.

$$a = 8eU / (mr_0 2\omega 2) \text{ and } q = 4eV_0 / (mr_0 2\omega 2)$$

(r_0) is half the distance between opposite rods.

$\omega = 2\pi f$; f is the frequency

At given U and V a specific mass/charge ion will reach the detector. When an unknown sample is analyzed, U and V are varied to get signal at detector. Once a U and V combination is giving signal, mass/charge can be calculated.

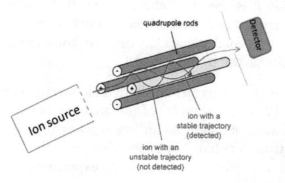

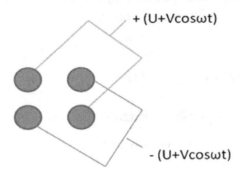

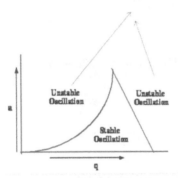

A simple setup for Quadruole analyzer and details of ion selection

Calculation of Molecular Mass of a lysozyme using following mass spectrometry data:

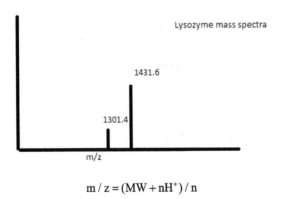

$$m/z = (MW + nH^+)/n$$

where, m/z = the mass–to–charge ratio

MW = the molecular mass ; n = the integer number of charges on the ions

H = the mass of a proton = 1.008 Da.

Usually the number of charges for each m/z signal is not known. However, two adjacent m/z signals in the series of multiply charged ions differ by one charge.

For example, if the ions appearing at m/z 1431.6 in the above figure have n charges, then the ions at m/z 1301.4 will have n+ 1 charge.

Thus,

$$1431.6 = (MW + nH^+)/n \text{ and } 1301.4 = [MW + (n+1)H^+]/(n+1)$$

After rearrangement of the equation to exclude the MW term:

$$n(1431.6) - nH^+ = (n+1)1301.4 - (n+1)H^+$$

and so:

$$n(1431.6) = n(1301.4) + 1301.4 - H^+$$

therefore:

$$n(1431.6 - 1301.4) = 1301.4 - H^+$$

and so:

$$n = (1301.4 - H^+)/(1431.6 - 1301.4)$$

Charges on the ions at m/z 1431.6 = 1300.4/130.2 = 10.

Once number of charge is calculated, molecular mass can easily be calculated as explained:

$$m/z = (MW + nH^+)/n$$

$$1431.6 = MW + (10 \times 1.008)/10$$

$$1431.6 \times 10 = MW + (10 \times 1.008)$$

$$MW = 14,316 - 10.08 = MW = 14,305.9 \text{ Da}$$

Try to Answer Following Question

A novel protein was investigated by various methods. Estimated molecular mass by size exclusion chromatography was 12000Da while by SDS–polyacrylamide gel electrophoresis molecular mass was found to be 13000Da. The sample was subjected to MALDI-TOF and following data obtained

m/z peak	773.9	825.5	884.3	952.3	1031.3
Abundance (%)	59	75	100	50	56

Solution:

- Abundance has no meaning in the question as the mass spectrometry is a qualitative method not quantitative.

- From mass analysis using othet investigations (size exclusion and SDS–poly acrylamide gel electrophoresis), it is clear that the molecular mass is between 12,000Da–13,000Da.

- For each m/z peak we need to calculate number of charge (protonation). For each m/z value we can calculate mass. Average of masses calculated from each m/z peak gives precise molecular mass of the protein.

m_1-1	$m_2 - m_1$	n_2	m_2-1	$M (Da) = n_2 (m_2-1)$	z
951.3	79.0	12.041	1030.3	12406.6	12
883.3	68.0	12.989	951.3	12357.1	13
824.5	25.8	14.022	883.3	12385.7	14
772.9	51.6	14.978	824.5	12349.9	15
			Mean M (Da)	12374.8	

Time of Flight

Time of flight (TOF) describes a variety of methods that measure the time that it takes for an object, particle or acoustic, electromagnetic or other wave to travel a distance through a medium. This measurement can be used for a time standard (such as an atomic fountain), as a way to measure velocity or path length through a given medium, or as a way to learn about the particle or medium (such as composition or flow rate). The traveling object may be detected directly (e.g., ion detector in mass spectrometry) or indirectly (e.g., light scattered from an object in laser doppler velocimetry).

- In electronics, the TOF method is used to estimate the electron mobility. Originally, it was designed for measurement of low-conductive thin films, later adjusted for common semiconductors. This experimental technique is used for metal-dielectric-metal structures as well as organic field-effect transistors. The excess charges are generated by application of the laser or voltage pulse.

- In time-of-flight mass spectrometry, ions are accelerated by an electrical field to the same kinetic energy with the velocity of the ion depending on the mass-to-charge ratio. Thus the time-of-flight is used to measure velocity, from which the mass-to-charge ratio can be determined. The time-of-flight of electrons is used to measure their kinetic energy.

- In near infrared spectroscopy, the TOF method is used to measure the media-dependent optical pathlength over a range of optical wavelengths, from which composition and properties of the media can be analyzed.

- In ultrasonic flow meter measurement, TOF is used to measure speed of signal propagation upstream and downstream of flow of a media, in order to estimate total flow velocity. This measurement is made in a collinear direction with the flow.

- In planar Doppler velocimetry (optical flow meter measurement), TOF measurements are made perpendicular to the flow by timing when individual particles cross two or more locations along the flow (collinear measurements would require generally high flow velocities and extremely narrow-band optical filters).

- In optical interferometry, the pathlength difference between sample and reference arms can be measured by TOF methods, such as frequency modulation followed by phase shift measurement or cross correlation of signals. Such methods are used in laser radar and laser tracker systems for medium-long range distance measurement.

- In Neutron time-of-flight scattering, a pulsed monochromatic neutron beam is scattered by a sample. The energy spectrum of the scattered neutrons is measured via time of flight.

- In kinematics, TOF is the duration in which a projectile is traveling through the air. Given the initial velocity u of a particle launched from the ground, the downward (i.e. gravitational) acceleration a, and the projectile's angle of projection θ (measured relative to the horizontal), then a simple rearrangement of the SUVAT equation

$$s = vt - \frac{1}{2}at^2$$

    results in this equation

$$t = \frac{2v\sin\theta}{a}$$

    for the time of flight of a projectile.

Time-of-flight Mass Spectrometry

Time-of-flight mass spectrometry (TOFMS) is a method of mass spectrometry in which ions are accelerated by an electric field of known strength. This acceleration results in an ion having the same kinetic energy as any other ion that has the same charge. The velocity of the ion depends on the mass-to-charge ratio. The time that it subsequently takes for the particle to reach a detector at a known distance is measured. This time will depend on the mass-to-charge ratio of the particle (heavier particles reach lower speeds). From this time and the known experimental parameters one can find the mass-to-charge ratio of the ion. The elapsed time from the instant a particle leaves a source to the instant it reaches a detector.

Shimadzu Ion Trap TOF

Ultrasonic and Optical Time-of-flight Flow Meters

An ultrasonic flow meter measures the velocity of a liquid or gas through a pipe using acoustic sensors. This has some advantages over other measurement techniques. The results are slightly affected by temperature, density or conductivity. Maintenance is inexpensive because there are no moving parts.

Ultrasonic flow meters come in three different types: transmission (contrapropagating transit time) flowmeters, reflection (Doppler) flowmeters, and open-channel flowmeters. Transit time flowmeters work by measuring the time difference between an ultrasonic pulse sent in the flow direction and an ultrasound pulse sent opposite the flow direction. Doppler flowmeters measure the doppler shift resulting in reflecting an ultrasonic beam off either small particles in the fluid, air bubbles in the fluid, or the flowing fluid's turbulence. Open channel flow meters measure upstream levels in front of flumes or weirs.

Optical time-of-flight sensors consist of two light beams projected into the fluid whose detection is either interrupted or instigated by the passage of small particles (which are assumed to be following the flow). This is not dissimilar from the optical beams used as safety devices in motorized garage doors or as triggers in alarm systems. The speed of the particles is calculated by knowing the spacing between the two beams. If there is only one detector, then the time difference can be measured via autocorrelation. If there are two detectors, one for each beam, then direction can also be known. Since the location of the beams is relatively easy to determine, the precision of the measurement depends primarily on how small the setup can be made. If the beams are too far apart, the flow could change substantially between them, thus the measurement becomes an average over that space. Moreover, multiple particles could reside between them at any given time, and this would corrupt the signal since the particles are indistinguishable.

For such a sensor to provide valid data, it must be small relative to the scale of the flow and the seeding density. MOEMS approaches yield extremely small packages, making such sensors applicable in a variety of situations.

High-precision Measurements in Physics

Usually the tube is praised for simplicity, but for precision measurements of charged low energy particles the electric and the magnetic field in the flight tube has to be controlled within 10 mV and 1 nT respectively.

The work function homogeneity of the tube can be controlled by a Kelvin probe. The magnetic field can be measured by a fluxgate compass. High frequencies are passively shielded and damped by radar absorbent material. To generate arbitrary low frequencies field the screen is parted into plates (overlapping and connected by capacitors) with bias voltage on each plate and a bias current on coil behind plate whose flux is closed by an outer core. In this way the tube can be configured to act as a weak achromatic quadrupole lens with an aperture with a grid and a delay line detector in the diffraction plane to do angle resolved measurements. Changing the field the angle of the field of view can be changed and a deflecting bias can be superimposed to scan through all angles.

When no delay line detector is used focusing the ions onto a detector can be accomplished through the use of two or three einzel lenses placed in the vacuum tube located between the ion source and the detector.

The sample should be immersed into the tube with holes and apertures for and against stray light to do magnetic experiments and to control the electrons from their start.

Sector Mass Spectrometer

A five sector mass spectrometer

A sector instrument is a general term for a class of mass spectrometer that uses a static electric or magnetic sector or some combination of the two (separately in space) as a mass analyzer. A popular combination of these sectors has been the BEB (magnetic-electric-magnetic). Most modern sector instruments are double-focusing instruments (first developed by A. Dempster, K. Bainbridge and J. Mattauch in 1936) in that they focus the ion beams both in direction and velocity.

Theory

The behavior of ions in a homogeneous, linear, static electric or magnetic field (separately) as is found in a sector instrument is simple. The physics are described by a single equation called the Lorentz force law. This equation is the fundamental equation of all mass spectrometric techniques and applies in non-linear, non-homogeneous cases too and is an important equation in the field of electrodynamics in general.

$$\mathbf{F} = q(\mathbf{E} + \mathbf{v} \times \mathbf{B}),$$

where E is the electric field strength, B is the magnetic field induction, q is the charge of the particle, **v** is its current velocity (expressed as a vector), and × is the cross product.

So the force on an ion in a linear homogeous electric field (an electric sector) is:

$$F = qE,$$

in the direction of the electric field, with positive ions and opposite that with negative ions.

Electric sector from a Finnigan MAT mass spectrometer (vacuum chamber housing removed)

The force is only dependent on the charge and electric field strength. The lighter ions will be deflected more and heavier ions less due to the difference in inertia and the ions will physically separate from each other in space into distinct beams of ions as they exit the electric sector.

And the force on an ion in a linear homogeneous magnetic field (a magnetic sector) is:

$$F = qvB,$$

perpendicular to both the magnetic field and the velocity vector of the ion itself, in the direction determined by the right-hand rule of cross products and the sign of the charge.

The force in the magnetic sector is complicated by the velocity dependence but with the right conditions (uniform velocity for example) ions of different masses will separate physically in space into different beams as with the electric sector.

Classic Geometries

These are some of the classic geometries from mass spectrographs which are often used to distinguish different types of sector arrangements, although most current instruments do not fit precisely into any of these categories as the designs have evolved further.

Bainbridge-Jordan

The sector instrument geometry consists of a 127.30° $\left(\frac{\pi}{\sqrt{2}}\right)$ electric sector without an initial drift length followed by a 60° magnetic sector with the same direction of curvature. Sometimes called a "Bainbridge mass spectrometer," this configuration is often used to determine isotopic masses. A beam of positive particles is produced from the isotope under study. The beam is subject to the combined action of perpendicular electric and magnetic fields. Since the forces due to these two fields are equal and opposite, the particles with a velocity given by

$$v = E / B$$

do not experience a resultant force; they pass freely through a slit, and are then subject to another magnetic field, transversing a semi-circular path and striking a photographic plate. The mass of the isotope is determined through subsequent calculation.

Mattauch-Herzog

The Mattauch-Herzog geometry consists of a 31.82° ($\pi/4\sqrt{2}$ radians) electric sector, a drift length which is followed by a 90° magnetic sector of opposite curvature direction. The entry of the ions sorted primarily by charge into the magnetic field produces an energy focussing effect and much higher transmission than a standard energy filter. This geometry is often used in applications with a high energy spread in the ions produced where sensitivity is nonetheless required, such as spark source mass spectrometry (SSMS) and secondary ion mass spectrometry (SIMS). The advantage of this geometry over the Nier-Johnson geometry is that the ions of different masses are all focused onto the same flat plane. This allows the use of a photographic plate or other flat detector array.

Nier-Johnson

The Nier-Johnson geometry consists of a 90° electric sector, a long intermediate drift length and a 60° magnetic sector of the same curvature direction.

Hinterberger-Konig

The Hinterberger-Konig geometry consists of a 42.43° electric sector, a long intermediate drift length and a 130° magnetic sector of the same curvature direction.

Takeshita

The Takeshita geometry consists of a 54.43° electric sector, and short drift length, a second electric sector of the same curvature direction followed by another drift length before a 180° magnetic sector of opposite curvature direction.

Matsuda

The Matsuda geometry consists of an 85° electric sector, a quadrupole lens and a 72.5° magnetic sector of the same curvature direction. This geometry is used in the SHRIMP and Panorama (gas source, high resolution, multicollector to measure isotopologues in geochemistry).

Quadrupole

A quadrupole or quadrapole is one of a sequence of configurations of—for example— electric charge or current, or gravitational mass that can exist in ideal form, but it is usually just part of a multipole expansion of a more complex structure reflecting various orders of complexity.

Mathematical Definition

The quadrupole moment tensor Q is a rank-two tensor (3x3 matrix) and is traceless (i.e. $Q_{xx} + Q_{yy} + Q_{zz} = 0$). The quadrupole moment tensor has thus 9 components, but because of the symmetry and zero-trace property, only 5 of these are independent.

For a discrete system of point charges (or masses in the case of a gravitational quadrupole), each with charge q_l (or mass m_l) and position $\vec{r}_l = (r_{xl}, r_{yl}, r_{zl})$ relative to the coordinate system origin, the components of the Q matrix are defined by:

$$Q_{ij} = \sum_l q_l (3 r_{il} r_{jl} - \|\vec{r}_l\|^2 \delta_{ij})$$

The indices i, j run over the Cartesian coordinates x, y, z and δ_{ij} is the Kronecker delta.

For a continuous system with charge density (or mass density) $\rho(x,y,z)$, the components of Q are defined by integral over the Cartesian space **r**:

$$Q_{ij} = \int \rho (3 r_i r_j - \|\vec{r}\|^2 \delta_{ij}) d^3 \mathbf{r}$$

As with any multipole moment, if a lower-order moment (monopole or dipole in this case) is non-zero, then the value of the quadrupole moment depends on the choice of the coordinate origin. For example, a dipole of two opposite-sign, same-strength point charges (which has no monopole moment) can have a nonzero quadrupole moment if the origin is shifted away from the center of the configuration (exactly between the two charges); or the quadrupole moment can be reduced to zero with the origin at the center. In contrast, if the monopole and dipole moments vanish, but the quadrupole moment does not (e.g., four same-strength charges, arranged in a square, with alternating signs), then the quadrupole moment is coordinate independent.

If each charge is the source of a "$1/r$" field, like the electric or gravitational field, the contribution to the field's potential from the quadrupole moment is:

$$V_q(\mathbf{R}) = \frac{k}{|\mathbf{R}|^3} \sum_{i,j} \frac{1}{2} Q_{ij} n_i n_j$$

where R is a vector with origin in the system of charges and **n** is the unit vector in the direction of R. Here, k is a constant that depends on the type of field, and the units being used. The factors n_i, n_j are components of the unit vector from the point of interest to the location of the quadrupole moment.

Electric Quadrupole

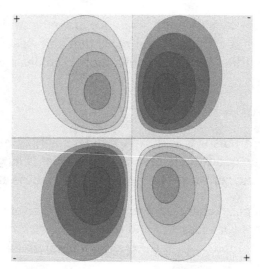

Contour plot of the equipotential surfaces of an electric quadrupole field

The simplest example of an electric quadrupole consists of alternating positive and negative charges, arranged on the corners of a square. The monopole moment (just the total charge) of this arrangement is zero. Similarly, the dipole moment is zero, regardless of the coordinate origin that has been chosen. But the quadrupole moment of the arrangement in the diagram cannot be reduced to zero, regardless of where we place the coordinate origin. The electric potential of an electric charge quadrupole is given by

An Overview of Mass Spectrometry

$$V_q(\mathbf{R}) = \frac{1}{4\pi\epsilon_0} \frac{1}{|\mathbf{R}|^3} \sum_{i,j} \frac{1}{2} Q_{ij} n_i n_j,$$

where ϵ_0 is the electric permittivity, and Q_{ij} follows the definition above.

Generalization: Higher Multipoles

An extreme generalization ("point octopole") would be: Eight alternating point charges at the eight corners of a parallelepiped, e.g. of a cube with edge length a. The "octopole moment" of this arrangement would correspond, in the "octopole limit" $\lim_{a\to 0;\, a^3 \cdot Q \to \text{const.}}$, to a nonzero diagonal tensor of order three. Still higher multipoles, e.g. of order 2^l, would be obtained by dipolar (quadrupolar, octopolar, ...) arrangements of point dipoles (quadrupoles, octopoles, ...), not point monopoles, of lower order, e.g. 2^{l-1}.

Magnetic Quadrupole

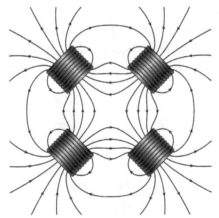
Coils producing a quadrupole field

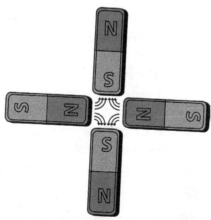
Schematic quadrupole magnet ("*four-pole*")

All known magnetic sources give dipole fields. However, to make a magnetic quadrupole it is possible to place four identical bar magnets perpendicular to each other such that the north pole of one is next to the south of the other. Such a configuration cancels the dipole moment and gives a quadrupole moment, and its field will decrease at large distances faster than that of a dipole.

An example of a magnetic quadrupole, involving permanent magnets, is depicted on the right. Electromagnets of similar conceptual design (called quadrupole magnets) are commonly used to focus beams of charged particles in particle accelerators and beam transport lines, a method known as strong focusing. The quadrupole-dipole intersect can be found by multiplying the spin of the unpaired nucleon by its parent atom. There are four steel pole tips, two opposing magnetic north poles and two opposing magnetic south poles. The steel is magnetized by a large electric current that flows in the coils of tubing wrapped around the poles.

A changing magnetic quadrupole moment produces electromagnetic radiation.

Gravitational Quadrupole

The mass quadrupole is analogous to the electric charge quadrupole, where the charge density is simply replaced by the mass density and a negative sign is added because the masses are always positive and the force is attractive. The gravitational potential is then expressed as:

$$V_q(\mathbf{R}) = -G \frac{1}{2} \frac{1}{|\mathbf{R}|^3} \sum_{i,j} Q_{ij} n_i n_j .$$

For example, because the Earth is rotating, it is oblate (flattened at the poles). This gives it a nonzero quadrupole moment. While the contribution to the Earth's gravitational field from this quadrupole is extremely important for artificial satellites close to Earth, it is less important for the Moon, because the $\frac{1}{|\mathbf{R}|^3}$ term falls quickly.

The mass quadrupole moment is also important in general relativity because, if it changes in time, it can produce gravitational radiation, similar to the electromagnetic radiation produced by oscillating electric or magnetic dipoles and higher multipoles. However, only quadrupole and higher moments can radiate gravitationally. The mass monopole represents the total mass-energy in a system, which is conserved—thus it gives off no radiation. Similarly, the mass dipole corresponds to the center of mass of a system and its first derivative represents momentum which is also a conserved quantity so the mass dipole also emits no radiation. The mass quadrupole, however, can change in time, and is the lowest-order contribution to gravitational radiation.

The simplest and most important example of a radiating system is a pair of mass points with equal masses orbiting each other on a circular orbit (an approximation to e.g. special case of binary black holes). Since the dipole moment is constant, we can for convenience place the coordinate origin right between the two points. Then the dipole moment will be zero, and if we also scale the coordinates so that the points are at unit distance from the center, in opposite direction, the system's quadrupole moment will then simply be

$$Q_{ij} = M(3x_i x_j - \delta_{ij})$$

where M is the mass of each point, and x_i are components of the (unit) position vector of one of the points. As they orbit, this **x**-vector will rotate, which means that it will have a nonzero first, and also the second time derivative (this is of course true regardless the choice of the coordinate system). Therefore the system will radiate gravitational waves. Energy lost in this way was first inferred in the changing period of the Hulse–Taylor binary pulsar, a pulsar in orbit with another neutron star of similar mass.

Just as electric charge and current multipoles contribute to the electromagnetic field, mass and mass-current multipoles contribute to the gravitational field in general relativity, causing the so-called "gravitomagnetic" effects. Changing mass-current multipoles can also give off gravitational radiation. However, contributions from the current multipoles will typically be much smaller than that of the mass quadrupole.

Principle of Mass Spectrometer

The first step in mass spectrometry involves generating charged particles in the gaseous form, which can be accelerated using a strong electric field and then analyzed by the detector.

In case of proteomic analysis, the proteins are digested with a suitable protease (usually trypsin) and the peptide fragments so obtained are ionized to generate charged particles. These charged particles are accelerated in vacuum under the presence of an external electric field. They are further fragmented (in tandem MS/MS) before finally being detected by the detector. The data is obtained in the form of relative abundance vs m/z ratio. Relative abundance refers to the abundance of that particular ion in the sample. Using the m/z ratio, an idea about the fragmentation pattern and hence the empirical formula of the molecule or peptide sequence of proteins can be established.

Components of Liquid Chromatography Mass Spectrometry

A Mass Spectrometer has three basic components – ionization source, mass analyzer and detector. Mass spectrometer is also coupled with HPLC system to facilitate easy separation of peptides, prior to their identification. These columns utilize the inherent properties of the peptides to separate them in solution, thereby increasing the resolution of the mass spectrometer.

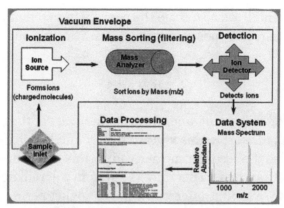

The basic components of mass spectrometer.

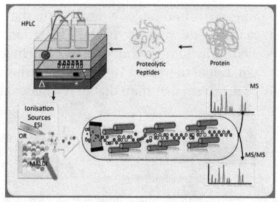

Schematic of a proteomic experiments. Protein is digested using trypsin and pre-fractionated using HPLC. MS device makes use of a combination of ion source and one or two mass analyzers

Chromatographic Unit

Protein mass spectrometry is a versatile tool for protein identification and quantification; however, it is subjected to various limitations. Since the detection is solely based on m/z ratio, it might so happen that two different peptide fragments are generated, having same m/z ratio. Assigning their true identity becomes a problem in this case. Also sometimes the highly abundant peptides overshadow the signals of low abundant peptides, which are clinically relevant from the point of view of biomarker discovery. Therefore, fractionation of complex mixtures of peptides into smaller fractions, which can be analyzed accurately and with high sensitivity, is required. The separation methodology using various types of HPLC techniques like RPLC (Reverse phase liquid chromatography), SXC (Strong Cation Exchanger), Affinity chromatography and Hydrophilic Interaction Liquid Chromatography (HILIC) is effective pre-fractionation strategy. RPLC and SXC are more commonly used to separate peptides.

RPLC: The columns for RPLC are highly hydrophobic C-18 columns, which strongly interact with the hydrophobic patches of the peptides. The major advantage of RPLC is that, the mobile phase used to elute peptides, is compatible with ESI. With varying degrees of hydrophobicity, the peptides elute at different time intervals and detected by MS.

SCX: The columns for SCX consist of strong cation exchangers like aliphatic sulfonic acid groups that are negatively charged in aqueous solution. Tryptic peptides containing positive charge are specifically exchanged using these columns. Using suitable solvents of high ionic strengths strong ionic interactions are broken. Ideally, working with only SCX yields very poor resolution of peptides. However, combining SCX with RPC yields better resolution, as two different properties of the peptides are being utilized in fractionating the peptides.

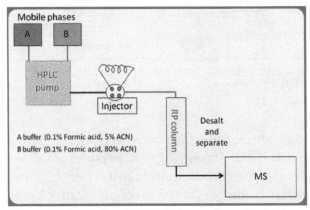

Scheme for RP-HPLC chromatography

Ionization Source

An ionization source ionizes the peptides so that they acquire a charge and hence can be accelerated in vacuum under the influence of an external electric field. The ionization source must be effective for all type of molecules (polar, non polar, non volatile etc.) and must be capable of ionizing the analyte without much degradation. Various ionization sources are described in Figure. Two most commonly used ionization sources MALDI and ESI will be discussed.

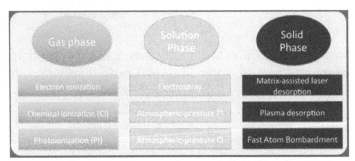

Different ionization sources: gas phase, solution phase & solid phase.

MALDI

The trypsin digested peptides are mixed with matrices (e.g. Cyanohydroxy cinnamic acid or cinapinic acid) and plated on MALDI plate. Short pulses of laser are inflicted upon the MALDI plate. The matrices absorb the energy from the laser and transmit it to the peptide fragments, which get ionized. Ideally singly charged ions are produced using MALDI ionization and these ions are accelerated through vacuum.

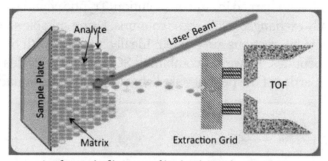

A schematic diagram of ionization using MALDI

ESI:

The need for softer ionization technique, to preserve peptide integrity, led to the emergence of Electrospray ionization (ESI). Trypsin digested peptides are allowed to pass through a very small capillary, the tips of which are maintained at very high voltage. As the peptides emerge out of the capillary in the form of fine droplets, the droplets get ionized. All type of charges are formed in ESI but it is the positively charged ions that are accelerated.

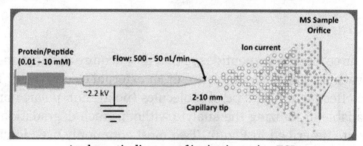

A schematic diagram of ionization using ESI.

Mass Analyzer

The mass analyzer represents the component of the mass spectrometer where the ionized peptides are separated according to their m/z ratio. Various types of mass analyzers are available, some of which are listed below.

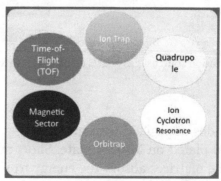

Different types of mass analyzers used for proteomics applications

TOF

Time of Flight involves accelerating the charged peptides through a long flight tube, maintained at vacuum. The different sized peptides by virtue of their difference in kinetic energy move at different rates and reach to the detector. Usually TOF is in association with MALDI and hence, singly charged peptides are accelerated through TOF. Thus heavier peptides take longer time to reach the detector than smaller peptides. The time required by a peptide to move across the entire flight tube is given by:

$$t = \left(\frac{m}{2qV_0}\right)^{1/2} L$$

Where
t = time-of-flight (s)
m = mass of the ion (kg)
q = charge on ion (C)
V_0 = accelerating potential (V)
L = length of flight tube (m)

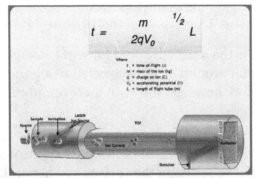

A time of flight (TOF) mass analyzer and its working principle. Ions are accelerated at different velocities depending on their m/z ratios. Ions of lower masses are accelerated to higher velocities and reach the detector first

Quadrupole / Triple Quadrupole (TQ)

A Quadrupole mass analyzer consists of a parallel set of four metallic rods, which are maintained at different potential difference, hence allowing particular ions to pass through them. A triple quadrupole consist of an additional collision cell in between two quadrupoles. The first quadrupole allows a particular set of ions to move into the collision cell, where that particular ion is fragmented into several ion pieces, all of which are then moved into the third quadrapole and hence the detector.

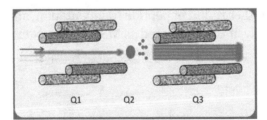

A typical scheme of Triple quadrupole mass analyzer.

ION TRAP

An ion trap mass analyzer consists of two sets of electrodes, the cap electrodes and the ring electrodes maintained at two different voltages. The ions that have a threshold m/z value more than the corresponding m/z value set by the voltage remain trapped in the ring electrode while other ions are allowed to pass into the cap electrode, where they are further fragmented and analyzed.

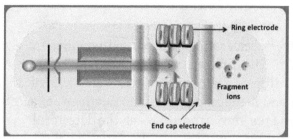

A typical scheme of Ion Trap mass analyzer.

Fourier Transform Ion Cyclotron Resonance (FT-ICR)

Ion cyclotron resonance is the most advanced form of mass analyzer where the ions of a particular m/z ratio orbit around a pole subjected to a uniform magnetic field. This gives the highest resolution; however, it is extremely complicate to handle.

Mass Detector

The mass detector is the final component of the mass spectrometer. It records the ion flux or the abundance of ions reaching the surface and hence the m/z ratio. Typically, a detector is an electron multiplier or an ion-to-photon converter (photo multiplier). This is because the amount ion reaching to the detector is extremely small to be detected and hence multiplication is required.

Tandem Mass Spectrometry

One of the most important applications of mass spectrometry is in protein sequencing. The Edman degradation method of protein sequencing is limited to a stretch of approximately 40 amino acids and also with the availability of a free N-terminal. Mass spectrometry based amino acid sequencing is hence a better option when it comes to protein identification.

Tandem MS refers to another round of peptide fragmentation, analysis and detection after the first round. Technically, it is represented as MSN, where N takes up an integer value.

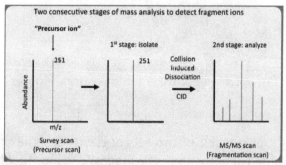

Tandem MS/MS - One of the peaks obtained is chosen and that particular peptide is allowed to move into the next mass analyzer where it is further fragmented and detected. Ideally, the different daughter peptides so obtained differ by one or more amino acid and hence from there, the sequence of the peptide can be determined

Hybrid Mass Spectrometer

Hybrid linear ion trap Fourier transform ion cyclotron resonance mass spectrometer

A hybrid mass spectrometer is a device for tandem mass spectrometry that consists of a combination of two or more m/z separation devices of different types.

Notation

The different m/z separation elements of a hybrid mass spectrometer can be represented by a shorthand notation. The symbol Q represents a quadrupole mass analyzer, q is a radio frequency collision quadrupole, TOF is a time-of-flight mass spectrometer, B is a magnetic sector and E is an electric sector.

Sector Quadrupole

A sector instrument can be combined with a collision quadrupole and quadrupole mass analyzer to form a hybrid instrument. A BEqQ configuration with a magnetic sector (B), electric sector (E), collision quadrupole (q) and m/z selection quadrupole (Q) have been constructed and an instrument with two electric sectors (BEEQ) has been described.

Quadrupole Time-of-flight

Hybrid quadrupole time-of-flight mass spectrometer.

A triple quadrupole mass spectrometer with the final quadrupole replaced by a time-of-flight device is known as a quadrupole time-of-flight instrument. Such an instrument can be represented as QqTOF.

Ion Trap Time-of-flight

In an ion trap instrument, ions are trapped in a quadrupole ion trap and then injected into the TOF. The trap can be 3-D or a linear trap.

Linear Ion Trap and Fourier Transform Mass Analyzers

A linear ion trap combined with a Fourier transform ion cyclotron resonance or orbitrap mass spectrometer is marketed by Thermo Scientific as the LTQ FT and LTQ Orbitrap, respectively.

Hybrid MS Configurations

Requirement of A Hybrid Ms Configuration

Before getting into the various hybrid MS configurations, it is essential to know why there was a requirement for combining different mass analyzers to give hybrid configurations. For example, the mass resolution of quadrapole was poor while that of TOF was medium, whereas the mass range of ion trap and quadrapole were low. On the other hand, FT-ICR which gave the best resolution and mass coverage was extremely complicated to operate. All these problems surfaced while studying various clinical samples, where the range of analytes may vary upto magnitudes of 108 (e.g. Human serum) and discovery of low molecular mass analytes require high sensitivity and mass accuracy.

A complex sample will contribute to various types of analytes. In the area of clinical proteomics, many proteins, having a wide concentration range may be present, and often these proteins act in regulating cellular pathways. On trypsin digestion the number of peptides generated may have overlapping properties, leading to their co-elution. Adding to the challenge is the need to detect very low-abundance species in the presence of highly abundant ones.

The technology of mass spectrometry was too versatile but individual mass analyzers had limitations hence scientist came up with the brilliant idea of making one weakness into the overall strength by combining two different weaklings, whose weakness are independent of each other. Thus was the origin of various combinations of hybrid mass analyzers. Tandem MS makes use of a combination of ion source and two mass analyzers (hybrid MS configurations), separated by a collision cell, in order to provide improved resolution of the fragment ions. The mass analyzers may either be the same or different. The first mass analyzer usually operates in a scanning mode in order to select only a particular ion, which is further fragmented and resolved in the second analyzer.

This can be used for protein sequencing studies.

Various Types of Hybrid Ms Configurations

Few hybrid MS configurations were developed to meet the needs of the complex samples.

a) TOF-TOF: Joining two time of flight tubes

b) Triple-Quadrapole: Joining two quadrapole at two ends with a collision cell in between

c) Q–TOF: Joining a quadrapole configuration with time of flight tube

d) LTQ–FTICR: Joining triple quadrapole with Ion cyclotron resonance

Comparison between Various Hybrid Configurations

For usage in day-to-day clinical proteomics, a configuration is necessary which gives high-throughput data as well as increased sensitivity and resolution. The reasons for such a criteria are already discussed above. All the configurations discussed have different advantages and limitations; hence to address different biological questions or to analyze different samples, different configurations may be adopted. For example, analyzing complex cerebrospinal fluids, where potential biomarkers may be present at attomolar range, Q–TOF or LTQ-Orbitrap may be beneficial, whereas for multiple reaction monitoring for biomarkers or drug metabolism, TQ may be advantageous. Thus the usage of a particular configuration is subject to the sample being processed, nonetheless all the hybrid MS configurations provide strong analytical platform for highly reproducible and accurate data required for proteomics applications.

Tandem Mass Spectrometry

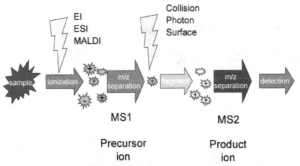

Schematic of tandem mass spectrometry

Tandem mass spectrometry, also known as MS/MS or MS^2, involves multiple steps of mass spectrometry selection, with some form of fragmentation occurring in between the stages. In a tandem mass spectrometer, ions are formed in the ion source and separated by mass-to-charge ratio in the first stage of mass spectrometry (MS1). Ions of a particular mass-to-charge ratio (precursor ions) are selected and fragment ions (product ions) are created by collision-induced dissociation, ion-molecule reaction, photodissociation, or other process. The resulting ions are then separated and detected in a second stage of mass spectrometry (MS2).

Instrumentation

For tandem mass spectrometry in space, the different elements are often noted in shorthand. Multiple stages of mass analysis separation can be accomplished with individual mass spectrometer elements separated in space or using a single mass spectrometer with the MS steps separated in time.

Tandem in Space

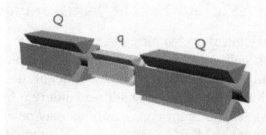

Triple quadrupole diagram; and example of tandem mass spectrometry in space.

In tandem mass spectrometry *in space*, the separation elements are physically separated and distinct, although there is a physical connection between the elements to maintain high vacuum. These elements can be sectors, transmission quadrupole, or time-of-flight. When using multiple quadrupoles, they can act as both mass analyzers and collision chambers.

- Q – Quadrupole mass analyzer
- q – Radio frequency collision quadrupole
- TOF – Time-of-flight mass analyzer
- B – Magnetic sector
- E – Electric sector

The notation can be combined to indicate various hybrid instrument, for example

- QqQ – Triple quadrupole mass spectrometer
- QTOF – Quadrupole time-of-flight mass spectrometer (also QqTOF)
- BEBE – Four-sector (reverse geometry) mass spectrometer

Tandem in Time

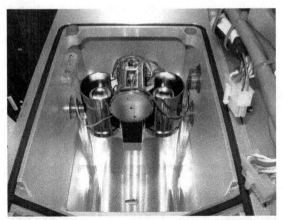

An ion trap mass spectrometer is an example of a tandem mass spectrometry in time instrument.

By doing tandem mass spectrometry *in time*, the separation is accomplished with ions trapped in the same place, with multiple separation steps taking place over time. A quadrupole ion trap or Fourier transform ion cyclotron resonance (FTICR) instrument can be used for such an analysis. Trapping instruments can perform multiple steps of analysis, which is sometimes referred to as MS^n (MS to the n). Often the number of steps, n, is not indicated, but occasionally the value is specified; for example MS^3 indicates three stages of separation. Tandem in time MS instruments do not use the modes described next, but typically collect all of the information from a precursor ion scan and a parent ion scan of the entire spectrum. Each instrumental configuration utilizes a unique mode of mass identification.

Tandem in Space MS/MS Modes

When tandem MS is performed with an in space design, the instrument must operate in one of a variety of modes. There are a number of different tandem MS/MS experimental setups and each mode has its own applications and provides different information. Tandem MS in space uses the coupling of two instrument components which measure the same mass spectrum range but with a controlled fractionation between them in space, while tandem MS in time involves the use of an ion trap.

There are four main scan experiments possible using MS/MS: precursor ion scan, product ion scan, neutral loss scan, and selected reaction monitoring.

For a precursor ion scan, the product ion is selected in the second mass analyzer, and the precursor masses are scanned in the first mass analyzer. Note that precursor ion is synonymous with parent ion and product ion with daughter ion; however the use of these anthropomorphic terms is discouraged.

In a product ion scan, a precursor ion is selected in the first stage, allowed to fragment and then all resultant masses are scanned in the second mass analyzer and detected in the detector that is positioned after the second mass analyzer. This experiment is commonly performed to identify transitions used for quantification by tandem MS.

In a neutral loss scan, the first mass analyzer scans all the masses. The second mass analyzer also scans, but at a set offset from the first mass analyzer. This offset corresponds to a neutral loss that is commonly observed for the class of compounds. In a constant-neutral-loss scan, all precursors that undergo the loss of a specified common neutral are monitored. To obtain this information, both mass analyzers are scanned simultaneously, but with a mass offset that correlates with the mass of the specified neutral. Similar to the precursor-ion scan, this technique is also useful in the selective identification of closely related class of compounds in a mixture.

In selected reaction monitoring, both mass analyzers are set to a selected mass. This mode is analogous to selected ion monitoring for MS experiments. A selective analysis mode, which can increase sensitivity.

Fragmentation

Fragmentation of gas-phase ions is essential to tandem mass spectrometry and occurs between different stages of mass analysis. There are many methods used to fragment the ions and these can result in different types of fragmentation and thus different information about the structure and composition of the molecule.

In-source Fragmentation

Often, the ionization process is sufficiently violent to leave the resulting ions with sufficient internal energy to fragment within the mass spectrometer. If the product ions persist in their non-equilibrium state for a moderate amount of time before auto-dissociation this process is called metastable fragmentation. Nozzle-skimmer fragmentation refers to the purposeful induction of in-source fragmentation by increasing the nozzle-skimmer potential on usually electrospray based instruments. Although in-source fragmentation allows for fragmentation analysis, it is not technically tandem mass spectrometry unless metastable ions are mass analyzed or selected before auto-dissociation and a second stage of analysis is performed on the resulting fragments. In-source fragmentation is often used in addition to tandem mass spectrometry (with post-source fragmentation) to allow for two steps of fragmentation in a pseudo MS^3-type of experiment.

Collision-induced Dissociation

Post-source fragmentation is most often what is being used in a tandem mass spectrometry experiment. Energy can also be added to the ions, which are usually already vibrationally excited, through post-source collisions with neutral atoms or molecules, the absorption of radiation, or the transfer or capture of an electron by a multiply charged ion. Collision-induced dissociation (CID), also called collisionally activated dissociation (CAD), involves the collision of an ion with a neutral atom or molecule in the gas phase and subsequent dissociation of the ion. For example, consider

$$AB^+ + M \rightarrow A + B^+ + M$$

where the ion AB^+ collides with the neutral species M and subsequently breaks apart. The details of this process are described by collision theory.

Higher-energy collisional dissociation (HCD) is a CID technique specific to orbitrap mass spectrometers in which fragmentation takes place external to the trap.

Electron Capture and Transfer Methods

The energy released when an electron is transferred to or captured by a multiply charged ion can induce fragmentation.

Electron capture dissociation

If an electron is added to a multiply charged positive ion, the Coulomb energy is liberated. Adding a free electron is called electron capture dissociation (ECD), and is represented by

$$[M+nH]^{n+} + e^- \rightarrow \left[[M+(n-1)H]^{(n-1)+}\right]^* \rightarrow \text{fragments}$$

for a multiply protonated molecule M.

Electron Transfer Dissociation

Adding an electron through an ion-ion reaction is called electron transfer dissociation (ETD). Similar to electron-capture dissociation, ETD induces fragmentation of cations (e.g. peptides or proteins) by transferring electrons to them. It was invented by Donald F. Hunt, Joshua Coon, John E. P. Syka and Jarrod Marto at the University of Virginia.

ETD does not use free electrons but employs radical anions (e.g. anthracene or azobenzene) for this purpose:

$$[M+nH]^{n+} + A^- \rightarrow \left[[M+(n-1)H]^{(n-1)+}\right]^* + A \rightarrow \text{fragments}$$

where A is the anion.

ETD cleaves randomly along the peptide backbone (c and z ions) while side chains and modifications such as phosphorylation are left intact. The technique only works well for higher charge state ions (z>2), however relative to collision-induced dissociation (CID), ETD is advantageous for the fragmentation of longer peptides or even entire proteins.

This makes the technique important for top-down proteomics. Much like ECD, ETD is effective for peptides with modifications such as phosphorylation.

Electron-transfer and higher-energy collision dissociation (EThcD) is a combination ETD and HCD where the peptide precursor is initially subjected to an ion/ion reaction with fluoranthene anions in a linear ion trap, which generates c- and z-ions. In the second step HCD all-ion fragmentation is applied to all ETD derived ions to generate b- and y- ions prior to final analysis in the orbitrap analyzer. This method employs dual fragmentation to generate ion- and thus data-rich MS/MS spectra for peptide sequencing and PTM localization.

Negative Electron Transfer Dissociation

Fragmentation can also occur with a deprotonated species, in which an electron is transferred from the specie to an cationic reagent in a negative electron transfer dissociation (NETD):

$$[M\text{-}n\text{H}]^{n-} + A^+ \rightarrow \left[[M\text{-}n\text{H}]^{(n+1)-}\right]^* + A \rightarrow \text{fragments}$$

Following this transfer event, the electron deficient anion undergoes internal rearrangement and fragments. NETD is the ion/ion analogue of electron-detachment dissociation (EDD).

NETD is compatible with fragmenting peptide and proteins along the backbone at the C_α-C bond. The resulting fragments are usually a·- and x-type product ions.

Electron-detachment Dissociation

Electron-detachment dissociation (EDD) is a method for fragmenting anionic species in mass spectrometry. It serves as a negative counter mode to electron capture dissociation. Negatively charged ions are activated by irradiation with electrons of moderate kinetic energy. The result is ejection of electrons from the parent ionic molecule, which causes dissociation via recombination.

Charge Transfer Dissociation

Reaction between positively charged peptides and cationic reagents, also known as charge transfer dissociation (CTD), has recently been demonstrated as an alternative high-energy fragmentation pathway for low-charge state (1+ or 2+) peptides. The proposed mechanism of CTD using helium cations as the reagent is:

$$[M+H]^{1+} + He^+ \rightarrow \left[[M+H]^{2+}\right]^* + He^0 \rightarrow \text{fragments}$$

Initial reports are that CTD causes backbone C_α-C bond cleavage of peptides and provides a·- and x-type product ions.

Photodissociation

The energy required for dissociation can be added by photon absorption, resulting in ion photodissociation and represented by

$$AB^+ + h\nu \rightarrow A + B^+$$

where $h\nu$ represents the photon absorbed by the ion. Ultraviolet lasers can be used, but can lead to excessive fragmentation of biomolecules.

Infrared multiphoton dissociation

Infrared photons will heat the ions and cause dissociation if enough of them are absorbed. This process is called infrared multiphoton dissociation (IRMPD) and is often accomplished with a carbon dioxide laser and an ion trapping mass spectrometer such as a FTMS.

Blackbody Infrared Radiative Dissociation

Blackbody radiation can be used for photodissociation in a technique known as blackbody infrared radiative dissociation (BIRD). In the BIRD method, the entire mass spectrometer vacuum chamber is heated to create infrared radiation. BIRD uses the light from black body radiation to thermally (vibrationally) excite the ions until a bond breaks. This is similar to infrared multiphoton dissociation with the exception of the source of radiation. This technique is most often used with Fourier transform ion cyclotron resonance mass spectrometers.

Surface Induced Dissociation

With surface-induced dissociation (SID), the fragmentation is a result of the collision of an ion with a surface under high vacuum.

Quantitative Proteomics

Quantitative proteomics is used to determine the relative or absolute amount of proteins in a sample. Several quantitative proteomics methods are based on tandem mass spectrometry. MS/MS has become a benchmark procedure for the structural elucidation of complex biomolecules.

Isobaric Tag for Relative and Absolute Quantitation

An isobaric tag for relative and absolute quantitation (iTRAQ) is a reagent for tandem mass spectrometry that is used to determine the amount of proteins from different sources in a single experiment. It uses stable isotope labeled molecules that can form a covalent bond with the N-terminus and side chain amines of proteins. The iTRAQ reagents are used to label peptides from different samples that are pooled and analyzed by liquid chromatography and tandem mass spectrometry. The fragmentation of the attached tag generates a low molecular mass reporter ion that can be used to relatively quantify the peptides and the proteins from which they originated.

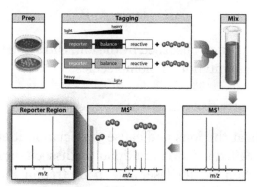

Isobaric labeling for tandem mass spectrometry: proteins are extracted from cells, digested, and labeled with tags of the same mass. When fragmented during MS/MS, the reporter ions show the relative amount of the peptides in the samples

Tandem Mass Tag

A tandem mass tag (TMT) is an isobaric mass tag chemical label used for protein quantification and identification. The tags contain four regions: mass reporter, cleavable linker, mass normalization, and protein reactive group.

Applications

Peptides

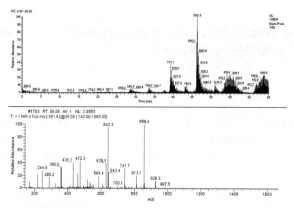

Chromatography trace (top) and tandem mass spectrum (bottom) of a peptide.

Tandem mass spectrometry can be used for protein sequencing. When intact proteins are introduced to a mass analyzer, this is called "top-down proteomics" and when proteins are digested into smaller peptides and subsequently introduced into the mass spectrometer, this is called "bottom-up proteomics". Shotgun proteomics is a variant of bottom up proteomics in which proteins in a mixture are digested prior to separation and tandem mass spectrometry.

Tandem mass spectrometry can produce a peptide sequence tag that can be used to identify a peptide in a protein database. A notation has been developed for indicating peptide fragments that arise from a tandem mass spectrum. Peptide fragment ions are indicated by a, b, or c if the charge is retained on the N-terminus and by x, y or z if the charge is maintained on the C-terminus. The subscript indicates the number of amino acid residues in the fragment. Superscripts are sometimes used to indicate neutral losses in addition to the backbone fragmentation, * for loss of ammonia and ° for loss of water. Although peptide backbone cleavage is the most useful for sequencing and peptide identification other fragment ions may be observed under high energy dissociation conditions. These include the side chain loss ions d, v, w and ammonium ions and additional sequence-specific fragment ions associated with particular amino acid residues.

Oligosaccharides

Oligosaccharides may be sequenced using tandem mass spectrometry in a similar manner to peptide sequencing. Fragmentation generally occurs on either side of the glycosidic bond (b, c, y and z ions) but also under more energetic conditions through the sugar ring structure in a cross-ring cleavage (x ions). Again trailing subscripts are used to indicate position of the cleavage along the chain. For cross ring cleavage ions the nature of the cross ring cleavage is indicated by preceding superscripts.

Oligonucleotides

Tandem mass spectrometry has been applied to DNA and RNA sequencing. A notation for gas-phase fragmentation of oligonucleotide ions has been proposed.

Newborn Screening

Newborn screening is the process of testing newborn babies for treatable genetic, endocrinologic, metabolic and hematologic diseases. The development of tandem mass spectrometry screening in the early 1990s led to a large expansion of potentially detectable congenital metabolic diseases that affect blood levels of organic acids.

Tandem Mass Spectrometry for Protein Identification

The use of mass spectrometry in field of protein chemistry has greatly revolutionized the field and contributed as one of the major factors for the emergence of proteomics. Mass spectrometry was initially used to identify the molecules based on their mass. However, due to advancements in mass analyzers and more sophisticated hybrid configurations, identifying and quantifying proteins has become one of the major applications of this technique. The use of mass spectrometry in determining the sequences of proteins and also de novo sequencing (where the genome sequences are not known) has gradually popularized due to the modification of existing mass spectrometers by adding another mass analyzer and detector in tandem, leading to the Tandem MS/MS analysis. We introduce the principle and importance of tandem MS/MS for protein identification.

Principle of Tandem Mass Spectrometry

Tandem mass spectrometry is a modified and advanced version of the conventional mass spectrometry. A mass spectrometer has three components: ionization source (generating ions), mass analyzer (filtering ions) and mass detector (generating the signal). Before we discuss new concepts it would be useful to refresh few concepts related to typical proteomics experiments, in-gel digestion and LC-MS/MS.

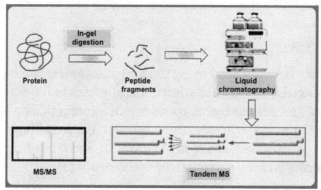

A typical LC-MS/MS set-up for proteomics applications.

Tandem mass spectrometry involves more than one additional mass analyzers and detectors, kept in tandem to the primary mass analyzer. The linear triple quadrapole (TQ) is an excellent example of Tandem Mass Spectrometry. A TQ has three components, Q1 operating at the scanning mode, i.e., allowing specific ions to move into Q2. Q2 is the chamber where CID (collision induced dissociation) takes place, and all the ions generating in Q2 move into Q3. Q3 or the third quadrapole operates in RF (radio frequency) mode, i.e., allowing all the ions, entering, to move towards the detector. You may also recall that the efficiency of TQ in terms of mass resolution and sensitivity and mass coverage was pretty poor. Therefore, hybrid mass configurations were developed having the basic TQ, associated to other high-end mass analyzers like TOF or Ion trap or FT – ICR. This increases the sensitivity and the resolution power of the approach and thus tandem MS/MS becomes useful for proteomics experiments.

Sequencing by Tandem Ms/Ms

b ION AND y ION

The first step for any mass spectrometry analysis is digestion of proteins by suitable proteases to yield small peptides, which can be detected easily and accurately by the mass spectrometer. The most common protease used is trypsin. Trypsin is an enzyme to digest peptides at the carboxyl terminal of lysine or arginine residues. Thus according to the amount of these basic amino acids in the protein, the number of peptides obtained post-tryptic digestion varies. The peptides so obtained are ionized and accelerated towards the mass analyzer. The peptides on their trajectory collide with other peptides and fragment further. The fragmentation mainly takes place at the CO-NH or amide bond, thereby generating daughter ions, which are positively charged either at the carboxyl side or at the amine side. If the positive charge is retained on the carboxyl side, it is known as the y ion, whereas if it is retained on the amine side, it is called the b ion.

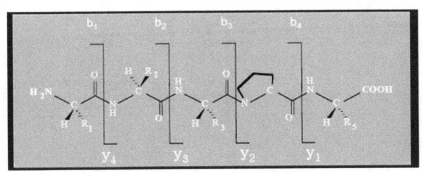

Peptide ion fragmentation and generation of b ion and y ions

Besides the b ions and y ions, several other types of ions are also generated, but those are usually not of significant importance, when it comes to sequencing purposes. The y ions are more intense than the b ions.

1. When the primary carbon and the amide carbon bond breaks, it gives rise to 'x' ions (positive charge retained on amide carbon) and 'a' ions (positive charge retained on primary carbon).

2. When the bond between amide nitrogen and primary carbon breaks, 'z' ions (positive charge retained on primary carbon) and 'c' ions (positive charge retained on amide nitrogen) are formed.

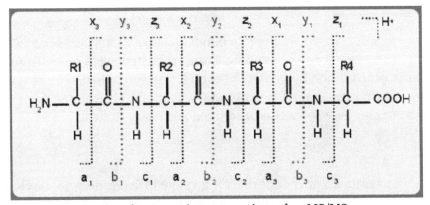

Various fragmentation patterns in tandem MS/MS

Protein Primary Structure

Diagram of protein structure, using PCNA as an example. (PDB: 1AXC)

Protein primary structure is the linear sequence of amino acids in a peptide or protein. By convention, the primary structure of a protein is reported starting from the amino-terminal (N) end to the carboxyl-terminal (C) end. Protein biosynthesis is most commonly performed by ribosomes in cells. Peptides can also be synthesised in the laboratory. Protein primary structures can be directly sequenced, or inferred from DNA sequences.

Formation

Biological

Amino acids are polymerised via peptide bonds to form a long backbone, with the different amino acid side chains protruding along it. In biological systems, proteins are produced during translation by a cell's ribosomes. Some organisms can also make short peptides by non-ribosomal peptide synthesis, which often use amino acids other than the standard 20, and may be cyclised, modified and cross-linked.

Chemical

Peptides can be synthesised chemically via a range of laboratory methods. Chemical methods typically synthesise peptides in the opposite order to biological protein synthesis (starting at the C-terminus).

Notation

Protein sequence is typically notated as a string of letters, listing the amino acids starting at the amino-terminal end through to the carboxyl-terminal end. Either a three letter code or single letter code can be used to represent the 20 naturally occurring amino acids, as well as mixtures or ambiguous amino acids (similar to nucleic acid notation).

Peptides can be directly sequenced, or inferred from DNA sequences. Large sequence databases now exist that collate known protein sequences.

20 natural amino acid notation		
Amino Acid	3-Letter	1-Letter
Alanine	Ala	A
Arginine	Arg	R
Asparagine	Asn	N
Aspartic acid	Asp	D
Cysteine	Cys	C
Glutamic acid	Glu	E
Glutamine	Gln	Q
Glycine	Gly	G
Histidine	His	H
Isoleucine	Ile	I
Leucine	Leu	L
Lysine	Lys	K
Methionine	Met	M
Phenylalanine	Phe	F
Proline	Pro	P
Serine	Ser	S
Threonine	Thr	T
Tryptophan	Trp	W
Tyrosine	Tyr	Y
Valine	Val	V

Ambiguous amino acid notation		
Symbol	Description	Residues represented
X	Any amino acid, or unknown	All
B	Aspartate or Asparagine	D, N
Z	Glutamate or Glutamine	E, Q
J	Leucine or Isoleucine	I, L
Φ	Hydrophobic	V, I, L, F, W, Y, M
Ω	Aromatic	F, W, Y, H
Ψ	Aliphatic	V, I, L, M

π	Small	P, G, A, S
ζ	Hydrophilic	S, T, H, N, Q, E, D, K, R
+	Positively charged	K, R, H
-	Negatively charged	D, E

Modification

In general, polypeptides are unbranched polymers, so their primary structure can often be specified by the sequence of amino acids along their backbone. However, proteins can become cross-linked, most commonly by disulfide bonds, and the primary structure also requires specifying the cross-linking atoms, e.g., specifying the cysteines involved in the protein's disulfide bonds. Other crosslinks include desmosine.

Isomerisation

The chiral centers of a polypeptide chain can undergo racemization. Although it does not change the sequence, it does affect the chemical properties of the sequence. In particular, the L-amino acids normally found in proteins can spontaneously isomerize at the C^α atom to form D-amino acids, which cannot be cleaved by most proteases. Additionally, proline can form stable trans-isomers at the peptide bond.

Posttranslational Modification

Finally, the protein can undergo a variety of posttranslational modifications, which are briefly summarized here.

The N-terminal amino group of a polypeptide can be modified covalently, e.g.,

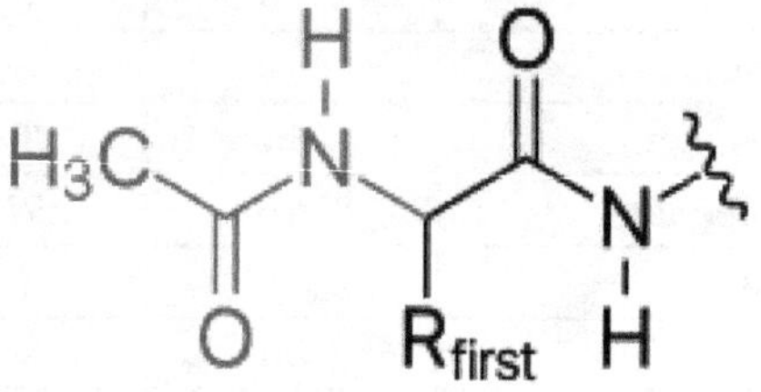

N-terminal acetylation

- acetylation $-C(=O)-CH_3$

   The positive charge on the N-terminal amino group may be eliminated by changing it to an acetyl group (N-terminal blocking).

- formylation $C(=O)H$

The N-terminal methionine usually found after translation has an N-terminus blocked with a formyl group. This formyl group (and sometimes the methionine residue itself, if followed by Gly or Ser) is removed by the enzyme deformylase.

- pyroglutamate

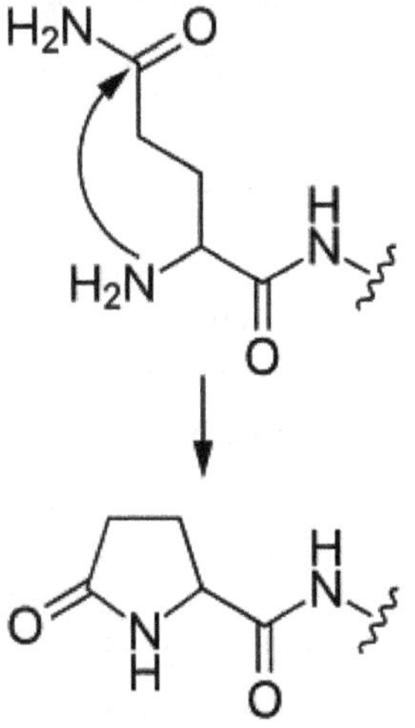

Formation of pyroglutamate from an N-terminal glutamine

An N-terminal glutamine can attack itself, forming a cyclic pyroglutamate group.

- myristoylation $-C(=O)-(CH_2)_{12}-CH_3$

Similar to acetylation. Instead of a simple methyl group, the myristoyl group has a tail of 14 hydrophobic carbons, which make it ideal for anchoring proteins to cellular membranes.

The C-terminal carboxylate group of a polypeptide can also be modified, e.g.,

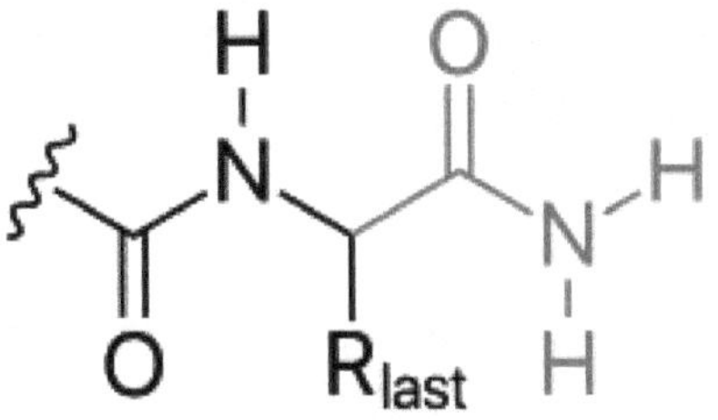

C-terminal amidation

- amidation

The C-terminus can also be blocked (thus, neutralizing its negative charge) by amidation.

- glycosyl phosphatidylinositol (GPI) attachment

Glycosyl phosphatidylinositol is a large, hydrophobic phospholipid prosthetic group that achors proteins to cellular membranes. It is attached to the polypeptide C-terminus through an amide linkage that then connects to ethanolamine, thence to sundry sugars and finally to the phosphatidylinositol lipid moiety.

Finally, the peptide side chains can also be modified covalently, e.g.,

- phosphorylation

    Aside from cleavage, phosphorylation is perhaps the most important chemical modification of proteins. A phosphate group can be attached to the sidechain hydroxyl group of serine, threonine and tyrosine residues, adding a negative charge at that site and producing an unnatural amino acid. Such reactions are catalyzed by kinases and the reverse reaction is catalyzed by phosphatases. The phosphorylated tyrosines are often used as "handles" by which proteins can bind to one another, whereas phosphorylation of Ser/Thr often induces conformational changes, presumably because of the introduced negative charge. The effects of phosphorylating Ser/Thr can sometimes be simulated by mutating the Ser/Thr residue to glutamate.

- glycosylation

    A catch-all name for a set of very common and very heterogeneous chemical modifications. Sugar moieties can be attached to the sidechain hydroxyl groups of Ser/Thr or to the sidechain amide groups of Asn. Such attachments can serve many functions, ranging from increasing solubility to complex recognition. All glycosylation can be blocked with certain inhibitors, such as tunicamycin.

- deamidation (succinimide formation)

    In this modification, an asparagine or aspartate side chain attacks the following peptide bond, forming a symmetrical succinimide intermediate. Hydrolysis of the intermediate produces either asparate or the β-amino acid, iso(Asp). For asparagine, either product results in the loss of the amide group, hence "deamidation".

- hydroxylation

    Proline residues may be hydroxylates at either of two atoms, as can lysine (at one atom). Hydroxyproline is a critical component of collagen, which becomes unstable upon its loss. The hydroxylation reaction is catalyzed by an enzyme that requires ascorbic acid (vitamin C), deficiencies in which lead to many connective-tissue diseases such as scurvy.

- methylation

    Several protein residues can be methylated, most notably the positive groups of lysine and arginine. Methylation at these sites is used to regulate the binding of proteins to nucleic acids. Lysine residues can be singly, doubly and even triply methylated. Methylation does *not* alter the positive charge on the side chain, however.

- acetylation

    Acetylation of the lysine amino groups is chemically analogous to the acetylation of the N-terminus. Functionally, however, the acetylation of lysine residues is used to regulate the binding of proteins to nucleic acids. The cancellation of the positive charge on the lysine weakens the electrostatic attraction for the (negatively charged) nucleic acids.

- sulfation

    Tyrosines may become sulfated on their O^η atom. Somewhat unusually, this modification occurs in the Golgi apparatus, not in the endoplasmic reticulum. Similar to phosphorylated tyrosines, sulfated tyrosines are used for specific recognition, e.g., in chemokine receptors on the cell surface. As with phosphorylation, sulfation adds a negative charge to a previously neutral site.

- prenylation and palmitoylation $-C(=O)-(CH_2)_{14}-CH_3$

    The hydrophobic isoprene (e.g., farnesyl, geranyl, and geranylgeranyl groups) and palmitoyl groups may be added to the S^γ atom of cysteine residues to anchor proteins to cellular membranes. Unlike the GPI and myritoyl anchors, these groups are not necessarily added at the termini.

- carboxylation

    A relatively rare modification that adds an extra carboxylate group (and, hence, a double negative charge) to a glutamate side chain, producing a Gla residue. This is used to strengthen the binding to "hard" metal ions such as calcium.

- ADP-ribosylation

    The large ADP-ribosyl group can be transferred to several types of side chains within proteins, with heterogeneous effects. This modification is a target for the powerful toxins of disparate bacteria, e.g., *Vibrio cholerae*, *Corynebacterium diphtheriae* and *Bordetella pertussis*.

- ubiquitination and SUMOylation

    Various full-length, folded proteins can be attached at their C-termini to the sidechain ammonium groups of lysines of other proteins. Ubiquitin is the most common of these, and usually signals that the ubiquitin-tagged protein should be degraded.

Most of the polypeptide modifications listed above occur *post-translationally*, i.e., after the protein has been synthesized on the ribosome, typically occurring in the endoplasmic reticulum, a subcellular organelle of the eukaryotic cell.

Many other chemical reactions (e.g., cyanylation) have been applied to proteins by chemists, although they are not found in biological systems.

Cleavage and Ligation

In addition to those listed above, the most important modification of primary structure is peptide cleavage (by chemical hydrolysis or by proteases). Proteins are often synthesized in an inactive precursor form; typically, an N-terminal or C-terminal segment blocks the active site of the protein, inhibiting its function. The protein is activated by cleaving off the inhibitory peptide.

Some proteins even have the power to cleave themselves. Typically, the hydroxyl group of a serine (rarely, threonine) or the thiol group of a cysteine residue will attack the carbonyl carbon of the preceding peptide bond, forming a tetrahedrally bonded intermediate [classified as a hydroxyoxazolidine (Ser/Thr) or hydroxythiazolidine (Cys) intermediate]. This intermediate tends to revert to the amide form, expelling the attacking group, since the amide form is usually favored by free energy, (presumably due to the strong resonance stabilization of the peptide group). However, additional molecular interactions may render the amide form less stable; the amino group is expelled instead, resulting in an ester (Ser/Thr) or thioester (Cys) bond in place of the peptide bond. This chemical reaction is called an N-O acyl shift.

The ester/thioester bond can be resolved in several ways:

- Simple hydrolysis will split the polypeptide chain, where the displaced amino group becomes the new N-terminus. This is seen in the maturation of glycosylasparaginase.

- A β-elimination reaction also splits the chain, but results in a pyruvoyl group at the new N-terminus. This pyruvoyl group may be used as a covalently attached catalytic cofactor in some enzymes, especially decarboxylases such as S-adenosylmethionine decarboxylase (SAMDC) that exploit the electron-withdrawing power of the pyruvoyl group.

- Intramolecular transesterification, resulting in a *branched* polypeptide. In inteins, the new ester bond is broken by an intramolecular attack by the soon-to-be C-terminal asparagine.

- Intermolecular transesterification can transfer a whole segment from one polypeptide to another, as is seen in the Hedgehog protein autoprocessing.

History

The proposal that proteins were linear chains of α-amino acids was made nearly simultaneously by two scientists at the same conference in 1902, the 74th meeting of the Society of German Scientists and Physicians, held in Karlsbad. Franz Hofmeister made the proposal in the morning, based on his observations of the biuret reaction in proteins. Hofmeister was followed a few hours later by Emil Fischer, who had amassed a wealth of chemical details supporting the peptide-bond model. For completeness, the proposal that proteins contained amide linkages was made as early as 1882 by the French chemist E. Grimaux.

Despite these data and later evidence that proteolytically digested proteins yielded only oligopeptides, the idea that proteins were linear, unbranched polymers of amino acids was not accepted immediately. Some well-respected scientists such as William Astbury doubted that covalent bonds were strong enough to hold such long molecules together; they feared that thermal agitations would shake such long molecules asunder. Hermann Staudinger faced similar prejudices in the 1920s when he argued that rubber was composed of macromolecules.

Thus, several alternative hypotheses arose. The colloidal protein hypothesis stated that proteins were colloidal assemblies of smaller molecules. This hypothesis was disproved in the 1920s by ultracentrifugation measurements by Theodor Svedberg that showed that proteins had a well-defined, reproducible molecular weight and by electrophoretic measurements by Arne Tiselius that indicated that proteins were single molecules. A second hypothesis, the cyclol hypothesis advanced by Dorothy Wrinch, proposed that the linear polypeptide underwent a chemical cyclol rearrangement C=O + HN → C(OH)-N that crosslinked its backbone amide groups, forming a two-dimensional *fabric*. Other primary structures of proteins were proposed by various researchers, such as the diketopiperazine model of Emil Abderhalden and the pyrrol/piperidine model of Troensegaard in 1942. Although never given much credence, these alternative models were finally disproved when Frederick Sanger successfully sequenced insulin and by the crystallographic determination of myoglobin and hemoglobin by Max Perutz and John Kendrew.

Primary Structure in Other Molecules

Any linear-chain heteropolymer can be said to have a "primary structure" by analogy to the usage of the term for proteins, but this usage is rare compared to the extremely common usage in reference to proteins. In RNA, which also has extensive secondary structure, the linear chain of bases is generally just referred to as the "sequence" as it is in DNA (which usually forms a linear double helix with little secondary structure). Other biological polymers such as polysaccharides can also be considered to have a primary structure, although the usage is not standard.

Relation to Secondary and Tertiary Structure

The primary structure of a biological polymer to a large extent determines the three-dimensional shape (tertiary structure). Protein sequence can be used to predict local features, such as segments of secondary structure, or trans-membrane regions. However, the complexity of protein folding currently prohibits predicting the tertiary structure of a protein from its sequence alone. Knowing the structure of a similar homologous sequence (for example a member of the same protein family) allows highly accurate prediction of the tertiary structure by homology modeling. If the full-length protein sequence is available, it is possible to estimate its general biophysical properties, such as its isoelectric point.

Sequence families are often determined by sequence clustering, and structural genomics projects aim to produce a set of representative structures to cover the sequence space of possible non-redundant sequences.

Factors Affecting Generation oF b IONS and y IONS

The generation of b ions and y ions is highly crucial for generation of protein sequences. The generation of b ions and y ions takes place under low collision energy, because of the difference in the electronegativity of carbon and nitrogen. The generation of different types of ions is quite natural, but strategies are developed to enrich the b and y ions.

Certain amino acids in the peptide backbone affect fragmentation pattern. For example, unusual fragmentation by loss of water may occur in serine and threonine, hydrogen sulphide loss in cysteine, ammonia loss from glutamic acid and aspartic acid, etc. Proline, an imino acid, affects the fragmentation pattern at both the levels of proteolysis by trypsin and at the level of generation of b and y ions. When proline is present at the carboxyl terminal after lysine or arginine, trypsin fails to cleave the peptides. This may be due to the cyclic nature of proline that inhibits the enzymatic action of trypsin. This problem can be mitigated using a different protease, like chymotrypsin, which cleaves at carboxyl terminal of aromatic amino acids. Cleavage near aspartic acid or glutamic acid residues generates intense ions, because of the excessive charge retained on the daughter peptide, which affects the ability to read correct sequences.

Deriving the Protein Sequence

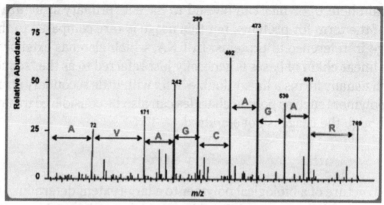

Deriving amino acid sequence from MS/MS spectrum

A typical fragmentation results into b ions and y ions, which differ in one amino acid, just like automated DNA sequencing using dideoxynucleotides, where the oligonucleotides differ from each other by one nucleotide. Identifying the amino acid from the difference in mass sequentially provides the sequence of the protein. The software (e.g., MASCOT) enables data analysis based on the mass of the various y ions and b ions. The difference between two successive y ions or b ions would yield the sequence of the amino acid.

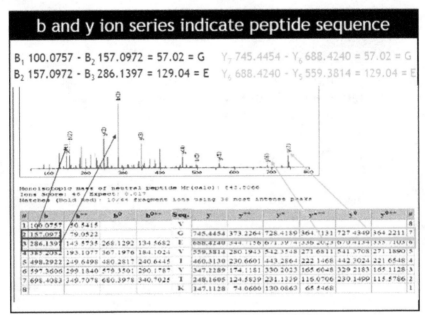

Generating sequences from b ions and y ions

De Novo Peptide Sequencing

In mass spectrometry, de novo peptide sequencing is the method in which a peptide amino acid sequence is determined from tandem mass spectrometry.

Knowing the amino acid sequence of peptides from a protein digest is essential to study the biological function of the protein. In the old days, this was accomplished by the Edman degradation procedure. Today, analysis by a tandem mass spectrometer is a more common method to solve the sequencing of peptides. Generally, there are two approaches: database search and de novo sequencing. Database search is a simple version as the mass spectra data of the unknown peptide is submitted and run to find a match with a known peptide sequence, the peptide with the highest matching score will be selected. This approach fails to recognize novel peptides since it can only match to existing sequences in the database. De novo sequencing is an assignment of fragment ions from a mass spectrum. Different algorithms are used for interpretation and most instruments come with de novo sequencing programs.

Peptide Fragmentation

Peptides are protonated in positive-ion mode. The proton initially locates at the N-terminus or a basic residue side chain, but because of the internal solvation, it can move along the backbone breaking at different sites which result in different fragments.

Three different types of backbone bonds can be broken to form peptide fragments: alkyl carbonyl (CHR-CO), peptide amide bond (CO-NH), and amino alkyl bond (NH-CHR).

Different Types of Fragment Ions

6 types of sequence ions in peptide fragmentation

When the backbone bonds cleave, six different types of sequence ions are formed as shown in Figure. The N-terminal charged fragment ions are classed as a, b or c, while the C-terminal charged ones are classed as x, y or z. The subscript n is the number of amino acid residues. The nomenclature was first proposed by Roepstorff and Fohlman, then Biemann modified it and this became the most widely accepted version.

Among these sequence ions, a, b and y-ions are the most common ion types, especially in the low-energy collision-induced dissociation (CID) mass spectrometers, since the peptide amide bone (CO-NH) is the most vulnerable and the loss of CO from b-ions.

Mass of b-ions = $\sum$ (residue masses) + 1 (H^+)

Mass of y-ions = $\sum$ (residue masses) + 19 (H_2O+H^+)

Mass of a-ions = mass of b-ions − 28 (CO)

Double backbone cleavage produces internal ions, acylium-type like $H_2N-CHR^2-CO-NH-CHR^3-CO+$ or immonium-type like $H_2N-CHR^2-CO-NH^+=CHR^3$. These ions are usually disturbance in the spectra.

Satellite ions in peptide fragmentation

Further cleavage happens under high-energy CID at the side chain of C-terminal residues, forming d_n, v_n, w_n-ions.

Fragmentation Rules

Most fragment ions are b- or y-ions. a-ions are also frequently seen by the loss of CO from b-ions.

Satellite ions (w_n, v_n, d_n-ions) are formed by high-energy CID.

Ser-, Thr-, Asp- and Glu-containing ions generate neutral molecular loss of water (-18).

Asn-, Gln-, Lys-, Arg-containing ions generate neutral molecular loss of ammonia (-17).

Neutral loss of ammonia from Arg leads to fragment ions (y-17) or (b-17) ions with higher abundant than their corresponding ions.

When C-terminus has a basic residue, the peptide generates (b_{n-1}+18) ion.

A complementary b-y ion pair can be observed in multiply charged ions spectra. For this b-y ion pair, the sum of their subscripts is equal to the total number of amino acid residues in the unknown peptide.

If the C-terminus is Arg or Lys, y_1-ion can be found in the spectrum to prove it.

Methods for Peptide Fragmentation

In low energy collision induced dissociation (CID), b- and y-ions are the main product ions. In addition, loss of ammonia (-17 Da) is observed in fragment with RKNQ amino acids in it. Loss of water (-18 Da) can be observed in fragment with STED amino acids in it. No satellite ions are shown in the spectra.

In high energy CID, all different types of fragment ions can be observed but no losses of ammonia or water.

In electron transfer dissociation (ETD) and electron capture dissociation (ECD), the predominant ions are c, y, z+1, z+2 and sometimes w ions.

For post source decay (PSD) in MALDI, a, b, y-ions are most common product ions.

Factors affecting fragmentation are the charge state (the higher charge state, the less energy is needed for fragmentation), mass of the peptide (the larger mass, the more energy is required), induced energy (higher energy leads to more fragmentation), primary amino acid sequence, mode of dissociation and collision gas.

Guidelines for Interpretation

Name	3-letter code	1-letter code	Residue Mass	Immonium ion	Related ions	Composition
Alanine	Ala	A	71.03711	44		C_3H_5NO
Arginine	Arg	R	156.10111	129	59,70,73,87,100,112	$C_6H_{12}N_4O$
Asparagine	Asn	N	114.04293	87	70	$C_4H_6N_2O_2$
Aspartic Acid	Asp	D	115.02694	88	70	$C_4H_5NO_3$
Cysteine	Cys	C	103.00919	76		C_3H_5NOS
Glutamic Acid	Glu	E	129.04259	102		$C_5H_7NO_3$
Glutamine	Gln	Q	128.05858	101	56,84,129	$C_5H_8N_2O_2$
Glycine	Gly	G	57.02146	30		C_2H_3NO
Histidine	His	H	137.05891	110	82,121,123,138,166	$C_6H_7N_3O$
Isoleucine	Ile	I	113.08406	86	44,72	$C_6H_{11}NO$
Leucine	Leu	L	113.08406	86	44,72	$C_6H_{11}NO$
Lysine	Lys	K	128.09496	101	70,84,112,129	$C_6H_{12}N_2O$
Methionine	Met	M	131.04049	104	61	C_5H_9NOS
Phenylalanine	Phe	F	147.06841	120	91	C_9H_9NO
Proline	Pro	P	97.05276	70		C_5H_7NO
Serine	Ser	S	87.03203	60		$C_3H_5NO_2$
Threonine	Thr	T	101.04768	74		$C_4H_7NO_2$
Tryptophan	Trp	W	186.07931	159	11,117,130,132,170,100	$C_{11}H_{10}N_2O$
Tyrosine	Tyr	Y	163.06333	136	91,107	$C_9H_9NO_2$
Valine	Val	V	99.06841	72	44,55,69	C_5H_9NO

Table. Mass of amino acid fragment ions

For interpretation, first, look for single amino acid immonium ions ($H_2N^+=CHR^2$). Corresponding immonium ions for amino acids are listed in Table. Ignore a few peaks at the high-mass end of the spectrum. They are ions that undergo neutral molecules losses (H_2O, NH_3, CO_2, HCOOH) from $[M+H]^+$ ions. Find mass differences at 28 Da since b-ions can form a-ions by loss of CO. Look for b_2-ions at low-mass end of the spectrum, which helps to identify y_{n-2}-ions too. Mass of b_2-ions are listed in Table, as well as single amino acids that have equal mass to b_2-ions. The mass of b_2-ion = mass of two amino acid residues + 1.

An Overview of Mass Spectrometry

	G	A	S	P	V	T	C	I/L	N	D	K/Q	E	M	H	F	R	Y	W
G	115																	
A	129	143																
S	145	159	175															
P	155	169	185	195														
V	157	171	187	197	199													
T	159	173	189	199	201	203												
C	161	175	191	201	203	205	207											
I/L	171	185	201	211	213	215	217	227										
N	172	186	202	212	214	216	218	228	229									
D	173	187	203	213	215	217	219	229	230	231								
K/Q	186	200	216	226	228	230	232	242	243	244	257							
E	187	201	217	227	229	231	233	243	244	245	258	259						
M	189	203	219	229	231	233	235	245	246	247	260	261	263					
H	195	209	225	235	237	239	241	251	252	253	266	267	269	275				
F[b]	205	219	235	245	247	249	251	261	262	263	276	277	279	285	295			
R	214	228	244	254	256	258	260	270	271	272	285	286	288	294	304	313		
Y	221	235	251	261	263	265	267	277	278	279	292	293	295	301	311	320	327	
W	244	258	274	284	286	288	290	300	301	302	315	316	318	324	334	343	350	373

GG=N=114; GA=K/Q=128; GV=R=156; GE=AD=SV=W=186.

Table. Mass of b2-ions in peptide fragmentaion

Identify a sequence ion series by the same mass difference, which matches one of the amino acid residue masses. For example, mass differences between a_n and a_{n-1}, b_n and b_{n-1}, c_n and c_{n-1} are the same. Identify y_{n-1}-ion at the high-mass end of the spectrum. Then continue to identify y_{n-2}, y_{n-3}... ions by matching mass differences with the amino acid residue masses. Look for the corresponding b-ions of the identified y-ions. The mass of b+y ions is the mass of the peptide +2 Da. After identifying the y-ion series and b-ion series, assign the amino acid sequence and check the mass. The other method is to identify b-ions first and then find the corresponding y-ions.

Algorithms and Software

Manual de novo sequencing is labor-intensive and time consuming. Usually algorithms or programs come with the mass spectrometer instrument are applied for the interpretation of spectra.

Development of de Novo Sequencing Algorithms

An old method is to list all possible peptides for the precursor ion in mass spectrum, and match the mass spectrum for each candidate to the experimental spectrum. The possible peptide that has the most similar spectrum will have the highest chance to be the right sequence. However, the number of possible peptides may be large. For example, a precursor peptide with a molecular weight of 774 has 21,909,046 possible peptides. Even though it is done in the computer, it takes a long time.

Another method is called "subsequencing", which instead of listing whole sequence of possible peptides, matches short sequences of peptides that represent only a part of the complete peptide. When sequences that highly match the fragment ions in the experimental spectrum are found, they are extended by residues one by one to find the best matching.

In the third method, graphical display of the data is applied, in which fragment ions that have the same mass differences of one amino acid residue are connected by lines. In this way, it is easier to get a clear image of ion series of the same type. This method could be helpful for manual de novo peptide sequencing, but doesn't work for high-throughput condition.

The fourth method, which is considered to be successful, is the graph theory. Applying graph theory in de novo peptide sequencing was first mentioned by Bartels. Peaks in the spectrum are transformed into vertices in a graph called "spectrum graph". If two vertices have the same mass difference of one or several amino acids, a directed edge will be applied. The SeqMS algorithm, Lutefisk algorithm, Sherenga algorithm are some examples of this type.

Software Packages

As described by Andreotti et al in 2012, Antilope is a combination of Lagrangian relaxation and an adaptation of Yen's k shortest paths. It is based on 'spectrum graph' method and contains different scoring functions, and can be comparable on the running time and accuracy to "the popular state-of-the-art programs" PepNovo and NovoHMM.

Grossmann et al presented AUDENS in 2005 as an automated de novo peptide sequencing tool cotaining a preprocessing module that can recognize signal peaks and noise peaks.

Lutefisk can solve de novo sequencing from CID mass spectra. In this algorithm, significant ions are first found, then determine the N- and C-terminal evidence list. Based on the sequence list, it generates complete sequences in spectra and scores them with the experimental spectrum. However, the result may include several sequence candidates that have only little difference, so it is hard to find the right peptide sequence. A second program, CIDentify, which is a modified version by Alex Taylor of Bill Pearson's FASTA algorithm, can be applied to distinguish those uncertain similar candidates.

Mo et al presented the MSNovo algorithm in 2007 and proved that it performed "better than existing de novo tools on multiple data sets". This algorithm can do de novo sequencing interpretation of LCQ, LTQ mass spectrometers and of singly, doubly, triply charged ions. Different from other algorithms, it applied a novel scoring function and use a mass array instead of a spectrum graph.

Fisher et al proposed the NovoHMM method of de novo sequencing. A hidden Markov model(HMM) is applied as a new way to solve de novo sequencing in a Bayesian framework. Instead of scoring for single symbols of the sequence, this method considers posterior probabilities for amino acids. In the paper, this method is proved to have better performance than other popular de novo peptide sequencing methods like PepNovo by a lot of example spectra.

PEAKS is a complete software package for the interpretation of peptide mass spectra. It contains de novo sequencing, database search, PTM identification, homology search and quantification in data analysis. Ma et al. described a new model and algorithm for de novo sequencing in PEAKS, and compared the performance with Lutefisk of several tryptic peptides of standard proteins, by the quadrupole time-of-flight(Q-TOF) mass spectrometer.

PepNovo is a high throughput de novo peptide sequencing tool and uses a probabilistic network as scoring method. It usually takes less than 0.2 seconds for interpretation of one spectrum. Described by Frank *et al*, PepNovo works better than several popular algorithms like Sherenga, PEAKS, Lutefisk. Now a new version PepNovo+ is available.

Chi *et al* presented pNovo+ in 2013 as a new de novo peptide sequencing tool by using complementary HCD and ETD tandem mass spectra. In this method, a component algorithm, pDAG, largely speeds up the acquisition time of peptide sequencing to 0.018s on average, which is three times as fast as the other popular de novo sequencing software.

As described by Jeong *et al*, compared with other do novo peptide sequencing tools, which works well on only certain types of spectra, UniNovo is a more universal tool that has a good performance on various types of spectra or spectral pairs like CID, ETD, HCD, CID/ETD, etc. It has a better accuracy than PepNovo+ or PEAKS. Moreover, it generates the error rate of the reported peptide sequences.

Ma published Novor in 2015 as a real-time de novo peptide sequencing engine. The tool is sought to improve the de novo speed by an order of magnitude and retain similar accuracy as other de novo tools in the market. On a Macbook Pro laptop, Novor has achieved more than 300 MS/MS spectra per second.

Pevtsov et al. compared the performance of the above five de novo sequencing algorithms: AUDENS, Lutefisk, NovoHMM, PepNovo, and PEAKS . QSTAR and LCQ mass spectrometer data were employed in the analysis, and evaluated by relative sequence distance (RSD) value, which was the similarity between de novo peptide sequencing and true peptide sequence calculated by a dynamic programming method. Results showed that all algorithms had better performance in QSTAR data than on LCQ data, while PEAKS as the best had a success rate of 49.7% in QSTAR data, and NovoHMM as the best had a success rate of 18.3% in LCQ data. The performance order in QSTAR data was PEAKS > Lutefisk, PepNovo > AUDENS, NovoHMM, and in LCQ data was NovoHMM > PepNovo, PEAKS > Lutefisk > AUDENS. Compared in a range of spectrum quality, PEAKS and NovoHMM also showed the best performance in both data among all 5 algorithms. PEAKS and NovoHMM had the best sensitivity in both QSTAR and LCQ data as well. However, no evaluated algorithms exceeded a 50% of exact identification for both data sets.

De-Novo Sequencing

Mass spectrometry is also used in sequencing proteins. The conventional approach of

Edman degradation has some of the limitations that are easily overcome by mass spectrometry. Firstly, Edman degradation is based on N-terminal sequencing of peptides, whereas there is no such constrains in mass spectrometry. Secondly, the efficiency of Edman degradation detection is up to 40 amino acids at a stretch, whereas the same for mass spectrometry in almost 100 amino acids at a stretch. In addition, the scan rate of mass spectrometry is much faster than the amino acid sequencer.

However, in conventional tandem mass spectrometry, the mass spectrometer, sequences a number of short peptide stretches and then by sequence homology from the database predicts the identity of the protein. But tandem mass spectrometry can also be used for generating de novo sequences of proteins, where the genome of the organism and hence the proteome is not available in the database. The overall principle of de novo sequencing is the same i.e., deriving the sequence of the peptide from the difference in mass of the peptides.

De novo sequencing finds its application in identifying novel splice variants and single nucleotide polymorphisms (SNPs), by the process of reverse genetics. The sequences of proteins so obtained can be reverse read as the genome sequence by software, after optimizing the degenerate genetic codes and codon biasness in those organisms.

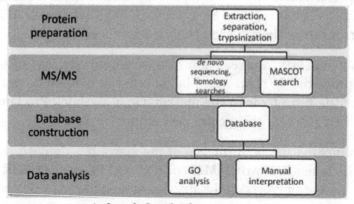

A typical work-flow for de novo sequencing

Challenges in Clinical Proteomics

Tandem MS/MS mainly finds its application in protein sequencing and de novo sequencing. It has become an integral part of the mass spectrometry. In clinical proteomics it is mainly used in biomarker discovery. But the area is still challenging, because validation is an important parameter for biomarker discovery, and the sample size also matters. The limitation in the field of clinical proteomics is partially because of instrumentations but mainly because of the pre-processing required for the samples and the large amount of sample that needs to be processed before it can be validated as biomarker. In general the tandem mass spectrometry has revolutionized proteomics by virtue of high-resolution mass spectrometers.

References

- Cotter, Robert J. (1994). Time-of-flight mass spectrometry. Columbus, OH: American Chemical Society. ISBN 0-8412-3474-4

- R.G. Kepler (1960). "Charge Carrier Production and Mobility in Anthracene Crystals". Phys. Rev. 119 (4): 1226. Bibcode:1960PhRv..119.1226K. doi:10.1103/PhysRev.119.1226

- Schoen, A.; Amy, J.W.; Ciupek, J.D.; Cooks, R.G.; Dobberstein, P.; Jung, G. (1985). "A hybrid BEQQ mass spectrometer". International Journal of Mass Spectrometry and Ion Processes. 65: 125–140. doi:10.1016/0168-1176(85)85059-X

- Hausman, Robert E.; Cooper, Geoffrey M. (2004). The cell: a molecular approach. Washington, D.C: ASM Press. p. 51. ISBN 0-87893-214-3

- M. Weis; J. Lin; D. Taguchi; T. Manaka; M. Iwamot (2009). "Analysis of Transient Currents in Organic Field Effect Transistor: The Time-of-Flight Method". J. Phys. Chem. C. 113 (43): 18459. doi:10.1021/jp908381b

- "Nier-Johnson geometry" (PDF). IUPAC Compendium of Chemical Terminology. IUPAC. 1997. Retrieved 2007-09-13

- Harrison, A. (1986). "A hybrid BEQQ mass spectrometer for studies in gaseous ion chemistry". International Journal of Mass Spectrometry and Ion Processes. 74: 13–31. doi:10.1016/0168-1176(86)85020-0

- Dass, Chhabil (2007). Fundamentals of contemporary mass spectrometry ([Online-Ausg.]. ed.). Hoboken, N.J.: Wiley-Interscience. pp. 317–322. ISBN 9780470118498. doi:10.1002/0470118490

- Burgoyne, Thomas W.; Gary M. Hieftje (1996). "An introduction to ion optics for the mass spectrograph" (abstract). Mass Spectrometry Reviews. 15 (4): 241–259. doi:10.1002/(SICI)1098-2787(1996)15:4<241::AID-MAS2>3.0.CO;2-I

- Edman, P.; Begg, G. (March 1967). "A Protein Sequenator". European Journal of Biochemistry. 1 (1): 80–91. PMID 6059350. doi:10.1111/j.1432-1033.1967.tb00047.x

- Dass, Chhabil (2001). Principles and practice of biological mass spectrometry. New York, NY [u.a.]: Wiley. ISBN 978-0-471-33053-0

- Weisstein, Eric. "Electric Quadrupole Moment". Eric Weisstein's World of Physics. Wolfram Research. Retrieved May 8, 2012

- Klemm, Alfred (1946). "The theory of a mass-spectrograph with double focus independent of mass". Zeitschrift für Naturforschung. 1: 137–41. Bibcode:1946ZNatA...1..137K. doi:10.1515/zna-1946-0306

- Yalcin, Talat; Csizmadia, Imre G.; Peterson, Michael R.; Harrison, Alex G. (March 1996). "The structure and fragmentation of B n (n≥3) ions in peptide spectra". Journal of the American Society for Mass Spectrometry. 7 (3): 233–242. doi:10.1016/1044-0305(95)00677-X

- McCloskey, edited by James A. (1990). Mass spectrometry. San Diego: Academic Press. pp. 886–887. ISBN 978-0121820947. CS1 maint: Extra text: authors list (link)

- De Laeter; J. & Kurz; M. D. (2006). "Alfred Nier and the sector field mass spectrometer". Journal of Mass Spectrometry. 41 (7): 847–854. doi:10.1002/jms.1057

- Dass, Chhabil (2007). Fundamentals of contemporary mass spectrometry ([Online-Ausg.]. ed.). Hoboken, N.J.: Wiley-Interscience. pp. 327–330. ISBN 9780470118498

Chromatography: A Comprehensive Study

Affinity chromatography falls under liquid chromatography and uses the method of reversible biological interaction. In theoretical terms, this technique can provide complete purification in a single step. Affinity chromatography is best understood in confluence with the major topics listed in the following chapter.

Chromatography

Chromatography is a laboratory technique for the separation of a mixture. The mixture is dissolved in a fluid called the *mobile phase,* which carries it through a structure holding another material called the *stationary phase*. The various constituents of the mixture travel at different speeds, causing them to separate. The separation is based on differential partitioning between the mobile and stationary phases. Subtle differences in a compound's partition coefficient result in differential retention on the stationary phase and thus changing the separation.

Pictured is a sophisticated gas chromatography system. This instrument records concentrations of acrylonitrile in the air at various points throughout the chemical laboratory

Chromatography may be preparative or analytical. The purpose of preparative chromatography is to separate the components of a mixture for later use, and is thus a form of purification. Analytical chromatography is done normally with smaller amounts of material and is for establishing the presence or measuring the relative proportions of analytes in a mixture. The two are not mutually exclusive.

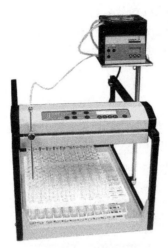
Automated fraction collector and sampler for chromatographic techniques

History

Thin layer chromatography is used to separate components of a plant extract, illustrating the experiment with plant pigments that gave chromatography its name

Chromatography was first employed in Russia by the Italian-born scientist Mikhail Tsvet in 1900. He continued to work with chromatography in the first decade of the 20th century, primarily for the separation of plant pigments such as chlorophyll, carotenes, and xanthophylls. Since these components have different colors (green, orange, and yellow, respectively) they gave the technique its name. New types of chromatography developed during the 1930s and 1940s made the technique useful for many separation processes.

Chromatography technique developed substantially as a result of the work of Archer John Porter Martin and Richard Laurence Millington Synge during the 1940s and

1950s, for which they won a Nobel prize. They established the principles and basic techniques of partition chromatography, and their work encouraged the rapid development of several chromatographic methods: paper chromatography, gas chromatography, and what would become known as high performance liquid chromatography. Since then, the technology has advanced rapidly. Researchers found that the main principles of Tsvet's chromatography could be applied in many different ways, resulting in the different varieties of chromatography described below. Advances are continually improving the technical performance of chromatography, allowing the separation of increasingly similar molecules.

Chromatography Terms

- The analyte is the substance to be separated during chromatography. It is also normally what is needed from the mixture.

- Analytical chromatography is used to determine the existence and possibly also the concentration of analyte(s) in a sample.

- A bonded phase is a stationary phase that is covalently bonded to the support particles or to the inside wall of the column tubing.

- A chromatogram is the visual output of the chromatograph. In the case of an optimal separation, different peaks or patterns on the chromatogram correspond to different components of the separated mixture.

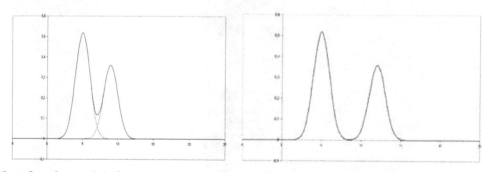

Plotted on the x-axis is the retention time and plotted on the y-axis a signal (for example obtained by a spectrophotometer, mass spectrometer or a variety of other detectors) corresponding to the response created by the analytes exiting the system. In the case of an optimal system the signal is proportional to the concentration of the specific analyte separated.

- A chromatograph is equipment that enables a sophisticated separation, e.g. gas chromatographic or liquid chromatographic separation.

- Chromatography is a physical method of separation that distributes components to separate between two phases, one stationary (stationary phase), the other (the mobile phase) moving in a definite direction.

- The eluate is the mobile phase leaving the column. This is also called effluent.

- The eluent is the solvent that carries the analyte.

- The eluite is the analyte, the eluted solute.

- An eluotropic series is a list of solvents ranked according to their eluting power.

- An immobilized phase is a stationary phase that is immobilized on the support particles, or on the inner wall of the column tubing.

- The mobile phase is the phase that moves in a definite direction. It may be a liquid (LC and Capillary Electrochromatography (CEC)), a gas (GC), or a supercritical fluid (supercritical-fluid chromatography, SFC). The mobile phase consists of the sample being separated/analyzed and the solvent that moves the sample through the column. In the case of HPLC the mobile phase consists of a non-polar solvent(s) such as hexane in normal phase or a polar solvent such as methanol in reverse phase chromatography and the sample being separated. The mobile phase moves through the chromatography column (the stationary phase) where the sample interacts with the stationary phase and is separated.

- Preparative chromatography is used to purify sufficient quantities of a substance for further use, rather than analysis.

- The retention time is the characteristic time it takes for a particular analyte to pass through the system (from the column inlet to the detector) under set conditions.

- The sample is the matter analyzed in chromatography. It may consist of a single component or it may be a mixture of components. When the sample is treated in the course of an analysis, the phase or the phases containing the analytes of interest is/are referred to as the sample whereas everything out of interest separated from the sample before or in the course of the analysis is referred to as waste.

- The solute refers to the sample components in partition chromatography.

- The solvent refers to any substance capable of solubilizing another substance, and especially the liquid mobile phase in liquid chromatography.

- The stationary phase is the substance fixed in place for the chromatography procedure. Examples include the silica layer in thin layer chromatography

- The detector refers to the instrument used for qualitative and quantitative detection of analytes after separation.

Chromatography is based on the concept of partition coefficient. Any solute partitions between two immiscible solvents. When we make one solvent immobile (by adsorption on a solid support matrix) and another mobile it results in most common applications of chromatography. If the matrix support, or stationary phase, is polar (e.g. paper, silica etc.) it is forward phase chromatography, and if it is non-polar (C-18) it is reverse phase.

Techniques by Chromatographic Bed Shape

Column Chromatography

Column chromatography is a separation technique in which the stationary bed is within a tube. The particles of the solid stationary phase or the support coated with a liquid stationary phase may fill the whole inside volume of the tube (packed column) or be concentrated on or along the inside tube wall leaving an open, unrestricted path for the mobile phase in the middle part of the tube (open tubular column). Differences in rates of movement through the medium are calculated to different retention times of the sample.

In 1978, W. Clark Still introduced a modified version of column chromatography called flash column chromatography (flash). The technique is very similar to the traditional column chromatography, except for that the solvent is driven through the column by applying positive pressure. This allowed most separations to be performed in less than 20 minutes, with improved separations compared to the old method. Modern flash chromatography systems are sold as pre-packed plastic cartridges, and the solvent is pumped through the cartridge. Systems may also be linked with detectors and fraction collectors providing automation. The introduction of gradient pumps resulted in quicker separations and less solvent usage.

In expanded bed adsorption, a fluidized bed is used, rather than a solid phase made by a packed bed. This allows omission of initial clearing steps such as centrifugation and filtration, for culture broths or slurries of broken cells.

Phosphocellulose chromatography utilizes the binding affinity of many DNA-binding proteins for phosphocellulose. The stronger a protein's interaction with DNA, the higher the salt concentration needed to elute that protein.

Planar Chromatography

Planar chromatography is a separation technique in which the stationary phase is present as or on a plane. The plane can be a paper, serving as such or impregnated by a substance as the stationary bed (paper chromatography) or a layer of solid particles spread on a support such as a glass plate (thin layer chromatography). Different compounds in the sample mixture travel different distances according to how strongly they interact with the stationary phase as compared to the mobile phase. The specific Retention factor (R_f) of each chemical can be used to aid in the identification of an unknown substance.

Paper Chromatography

Paper chromatography is a technique that involves placing a small dot or line of sample solution onto a strip of *chromatography paper*. The paper is placed in a container with a shallow layer of solvent and sealed. As the solvent rises through the paper, it meets the sample mixture, which starts to travel up the paper with the solvent. This paper is made of cellulose, a polar substance, and the compounds within the mixture travel farther if they are non-polar. More polar substances bond with the cellulose paper more quickly, and therefore do not travel as far.

Thin Layer Chromatography (TLC)

Thin layer chromatography (TLC) is a widely employed laboratory technique use to separate different biochemicals on the basis of their size and is similar to paper chromatography. However, instead of using a stationary phase of paper, it involves a stationary phase of a thin layer of adsorbent like silica gel, alumina, or cellulose on a flat, inert substrate. Compared to paper, it has the advantage of faster runs, better separations, and the choice between different adsorbents. For even better resolution and to allow for quantification, high-performance TLC can be used. An older popular use had been to differentiate chromosomes by observing distance in gel (separation of was a separate step).

Displacement Chromatography

The basic principle of displacement chromatography is: A molecule with a high affinity for the chromatography matrix (the displacer) competes effectively for binding sites, and thus displace all molecules with lesser affinities. There are distinct differences between displacement and elution chromatography. In elution mode, substances typically emerge from a column in narrow, Gaussian peaks. Wide separation of peaks, preferably to baseline, is desired for maximum purification. The speed at which any component of a mixture travels down the column in elution mode depends on many factors. But for two substances to travel at different speeds, and thereby be resolved, there must be substantial differences in some interaction between the biomolecules and the chromatography matrix. Operating parameters are adjusted to maximize the effect of this difference. In many cases, baseline separation of the peaks can be achieved only with gradient elution and low column loadings. Thus, two drawbacks to elution mode chromatography, especially at the preparative scale, are operational complexity, due to gradient solvent pumping, and low throughput, due to low column loadings. Displacement chromatography has advantages over elution chromatography in that components are resolved into consecutive zones of pure substances rather than "peaks". Because the process takes advantage of the nonlinearity of the isotherms, a larger column feed can be separated on a given column with the purified components recovered at significantly higher concentrations.

Techniques by Physical State of Mobile Phase

Gas Chromatography

Gas chromatography (GC), also sometimes known as gas-liquid chromatography, (GLC), is a separation technique in which the mobile phase is a gas. Gas chromatographic separation is always carried out in a column, which is typically "packed" or "capillary". Packed columns are the routine work horses of gas chromatography, being cheaper and easier to use and often giving adequate performance. Capillary columns generally give far superior resolution and although more expensive are becoming widely used, especially for complex mixtures. Both types of column are made from non-adsorbent and chemically inert materials. Stainless steel and glass are the usual materials for packed columns and quartz or fused silica for capillary columns.

Gas chromatography is based on a partition equilibrium of analyte between a solid or viscous liquid stationary phase (often a liquid silicone-based material) and a mobile gas (most often helium). The stationary phase is adhered to the inside of a small-diameter (commonly 0.53 – 0.18mm inside diameter) glass or fused-silica tube (a capillary column) or a solid matrix inside a larger metal tube (a packed column). It is widely used in analytical chemistry; though the high temperatures used in GC make it unsuitable for high molecular weight biopolymers or proteins (heat denatures them), frequently encountered in biochemistry, it is well suited for use in the petrochemical, environmental monitoring and remediation, and industrial chemical fields. It is also used extensively in chemistry research.

Liquid Chromatography

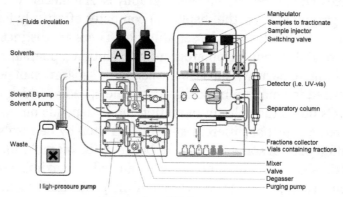

Preparative HPLC apparatus

Liquid chromatography (LC) is a separation technique in which the mobile phase is a liquid. It can be carried out either in a column or a plane. Present day liquid chromatography that generally utilizes very small packing particles and a relatively high pressure is referred to as high performance liquid chromatography (HPLC).

In HPLC the sample is forced by a liquid at high pressure (the mobile phase) through a column that is packed with a stationary phase composed of irregularly or spherically shaped particles, a porous monolithic layer, or a porous membrane. HPLC is historically divided into two different sub-classes based on the polarity of the mobile and stationary phases. Methods in which the stationary phase is more polar than the mobile phase (e.g., toluene as the mobile phase, silica as the stationary phase) are termed normal phase liquid chromatography (NPLC) and the opposite (e.g., water-methanol mixture as the mobile phase and C18 (octadecylsilyl) as the stationary phase) is termed reversed phase liquid chromatography (RPLC).

Specific techniques under this broad heading are listed below.

Affinity Chromatography

Affinity chromatography is based on selective non-covalent interaction between an analyte and specific molecules. It is very specific, but not very robust. It is often used in biochemistry in the purification of proteins bound to tags. These fusion proteins are labeled with compounds such as His-tags, biotin or antigens, which bind to the stationary phase specifically. After purification, some of these tags are usually removed and the pure protein is obtained.

Affinity chromatography often utilizes a biomolecule's affinity for a metal (Zn, Cu, Fe, etc.). Columns are often manually prepared. Traditional affinity columns are used as a preparative step to flush out unwanted biomolecules.

However, HPLC techniques exist that do utilize affinity chromatography properties. Immobilized Metal Affinity Chromatography (IMAC) is useful to separate aforementioned molecules based on the relative affinity for the metal (I.e. Dionex IMAC). Often these columns can be loaded with different metals to create a column with a targeted affinity.

Supercritical Fluid Chromatography

Supercritical fluid chromatography is a separation technique in which the mobile phase is a fluid above and relatively close to its critical temperature and pressure.

Techniques by Separation Mechanism

Ion Exchange Chromatography

Ion exchange chromatography (usually referred to as ion chromatography) uses an ion exchange mechanism to separate analytes based on their respective charges. It is usually performed in columns but can also be useful in planar mode. Ion exchange chromatography uses a charged stationary phase to separate charged compounds including anions, cations, amino acids, peptides, and proteins. In conventional methods the stationary phase is an ion exchange resin that carries charged functional groups that interact with oppositely charged groups of the compound to retain. Ion exchange chromatography is commonly used to purify proteins using FPLC.

Size-exclusion Chromatography

Size-exclusion chromatography (SEC) is also known as gel permeation chromatography (GPC) or gel filtration chromatography and separates molecules according to their size (or more accurately according to their hydrodynamic diameter or hydrodynamic volume). Smaller molecules are able to enter the pores of the media and, therefore, molecules are trapped and removed from the flow of the mobile phase. The average residence time in the pores depends upon the effective size of the analyte molecules. However, molecules that are larger than the average pore size of the packing are excluded and thus suffer essentially no retention; such species are the first to be eluted. It is generally a low-resolution chromatography technique and thus it is often reserved for the final, "polishing" step of a purification. It is also useful for determining the tertiary structure and quaternary structure of purified proteins, especially since it can be carried out under native solution conditions.

Expanded Bed Adsorption Chromatographic Separation

An expanded bed chromatographic adsorption (EBA) column for a biochemical separation process comprises a pressure equalization liquid distributor having a self-cleaning function below a porous blocking sieve plate at the bottom of the expanded bed, an upper part nozzle assembly having a backflush cleaning function at the top of the expanded bed, a better distribution of the feedstock liquor added into the expanded bed ensuring that the fluid passed through the expanded bed layer displays a state of piston flow. The expanded bed layer displays a state of piston flow. The expanded bed chromatographic separation column has advantages of increasing the separation efficiency of the expanded bed.

Expanded-bed adsorption (EBA) chromatography is a convenient and effective technique for the capture of proteins directly from unclarified crude sample. In EBA chromatography, the settled bed is first expanded by upward flow of equilibration buffer. The crude feed, a mixture of soluble proteins, contaminants, cells, and cell debris, is then passed upward through the expanded bed. Target proteins are captured on the adsorbent, while particulates and contaminants pass through. A change to elution buffer while maintaining upward flow results in desorption of the target protein in expanded-bed mode. Alternatively, if the flow is reversed, the adsorbed particles will quickly settle and the proteins can be desorbed by an elution buffer. The mode used for elution (expanded-bed versus settled-bed) depends on the characteristics of the feed. After elution, the adsorbent is cleaned with a predefined cleaning-in-place (CIP) solution, with cleaning followed by either column regeneration (for further use) or storage.

Special Techniques

Reversed-phase Chromatography

Reversed-phase chromatography (RPC) is any liquid chromatography procedure in which the mobile phase is significantly more polar than the stationary phase. It is so named because in normal-phase liquid chromatography, the mobile phase is significantly less polar than the stationary phase. Hydrophobic molecules in the mobile phase tend to adsorb to the relatively hydrophobic stationary phase. Hydrophilic molecules in the mobile phase will tend to elute first. Separating columns typically comprise a C8 or C18 carbon-chain bonded to a silica particle substrate.

Hydrophobic Interaction Chromatography

Hydrophobic interactions between proteins and the chromatographic matrix can be exploited to purify proteins. In hydrophobic interaction chromatography the matrix material is lightly substituted with hydrophobic groups. These groups can range from methyl, ethyl, propyl, octyl, or phenyl groups. At high salt concentrations, non-polar sidechains on the surface on proteins "interact" with the hydrophobic groups; that is, both types of groups are excluded by the polar solvent (hydrophobic effects are augmented by increased ionic strength). Thus, the sample is applied to the column in a buffer which is highly polar. The eluant is typically an aqueous buffer with decreasing salt concentrations, increasing concentrations of detergent (which disrupts hydrophobic interactions), or changes in pH.

In general, Hydrophobic Interaction Chromatography (HIC) is advantageous if the sample is sensitive to pH change or harsh solvents typically used in other types of chromatography but not high salt concentrations. Commonly, it is the amount of salt in the buffer which is varied. In 2012, Müller and Franzreb described the effects of temperature on HIC using Bovine Serum Albumin (BSA) with four different types of hydrophobic resin. The study altered temperature as to effect the binding affinity of BSA onto the matrix. It was concluded that cycling temperature from 50 degrees to 10 degrees would not be adequate to effectively wash all BSA from the matrix but could be very effective if the column would only be used a few times. Using temperature to effect change allows labs to cut costs on buying salt and saves money.

Two-dimensional chromatograph GCxGC-TOFMS at Chemical Faculty of GUT Gdańsk, Poland, 2016

If high salt concentrations along with temperature fluctuations want to be avoided you can use a more hydrophobic to compete with your sample to elute it. [source] This so-called salt independent method of HIC showed a direct isolation of Human Immunoglobulin G (IgG) from serum with satisfactory yield and used Beta-cyclodextrin as a competitor to displace IgG from the matrix. This largely opens up the possibility of using HIC with samples which are salt sensitive as we know high salt concentrations precipitate proteins.

Two-dimensional Chromatography

In some cases, the chemistry within a given column can be insufficient to separate some analytes. It is possible to direct a series of unresolved peaks onto a second column with different physico-chemical (Chemical classification) properties. Since the mechanism of retention on this new solid support is different from the first dimensional separation, it can be possible to separate compounds that are indistinguishable by one-dimensional chromatography. The sample is spotted at one corner of a square plate, developed, air-dried, then rotated by 90° and usually redeveloped in a second solvent system.

Simulated Moving-bed Chromatography

The simulated moving bed (SMB) technique is a variant of high performance liquid chromatography; it is used to separate particles and/or chemical compounds that would be difficult or impossible to resolve otherwise. This increased separation is brought about by a valve-and-column arrangement that is used to lengthen the stationary phase indefinitely. In the moving bed technique of preparative chromatography the feed entry and the analyte recovery are simultaneous and continuous, but because of practical difficulties with a continuously moving bed, simulated moving bed technique was proposed. In the simulated moving bed technique instead of moving the bed, the sample inlet and the analyte exit positions are moved continuously, giving the impression of a moving bed. True moving bed chromatography (TMBC) is only a theoretical concept. Its simulation, SMBC is achieved by the use of a multiplicity of columns in series and a complex valve arrangement, which provides for sample and solvent feed, and also analyte and waste takeoff at appropriate locations of any column, whereby it allows switching at regular intervals the sample entry in one direction, the solvent entry in the opposite direction, whilst changing the analyte and waste takeoff positions appropriately as well.

Pyrolysis Gas Chromatography

Pyrolysis gas chromatography mass spectrometry is a method of chemical analysis in which the sample is heated to decomposition to produce smaller molecules that are separated by gas chromatography and detected using mass spectrometry.

Pyrolysis is the thermal decomposition of materials in an inert atmosphere or a vacuum. The sample is put into direct contact with a platinum wire, or placed in a quartz sample tube, and rapidly heated to 600–1000 °C. Depending on the application even higher temperatures are used. Three different heating techniques are used in actual pyrolyzers: Isothermal furnace, inductive heating (Curie Point filament), and resistive heating using platinum filaments. Large molecules cleave at their weakest points and produce smaller, more volatile fragments. These fragments can be separated by gas chromatography. Pyrolysis GC chromatograms are typically complex because a wide range of different decomposition products is formed. The data can either be used as fingerprint to prove material identity or the GC/MS data is used to identify individual fragments to obtain structural information. To increase the volatility of polar fragments, various methylating reagents can be added to a sample before pyrolysis.

Besides the usage of dedicated pyrolyzers, pyrolysis GC of solid and liquid samples can be performed directly inside Programmable Temperature Vaporizer (PTV) injectors that provide quick heating (up to 30 °C/s) and high maximum temperatures of 600–650 °C. This is sufficient for some pyrolysis applications. The main advantage is that no dedicated instrument has to be purchased and pyrolysis can be performed as part of routine GC analysis. In this case quartz GC inlet liners have to be used. Quantitative data can be acquired, and good results of derivatization inside the PTV injector are published as well.

Fast Protein Liquid Chromatography

Fast protein liquid chromatography (FPLC), is a form of liquid chromatography that is often used to analyze or purify mixtures of proteins. As in other forms of chromatography, separation is possible because the different components of a mixture have different affinities for two materials, a moving fluid (the "mobile phase") and a porous solid (the stationary phase). In FPLC the mobile phase is an aqueous solution, or "buffer". The buffer flow rate is controlled by a positive-displacement pump and is normally kept constant, while the composition of the buffer can be varied by drawing fluids in different proportions from two or more external reservoirs. The stationary phase is a resin composed of beads, usually of cross-linked agarose, packed into a cylindrical glass or plastic column. FPLC resins are available in a wide range of bead sizes and surface ligands depending on the application.

Countercurrent Chromatography

Countercurrent chromatography (CCC) is a type of liquid-liquid chromatography, where both the stationary and mobile phases are liquids. The operating principle of CCC equipment requires a column consisting of an open tube coiled around a bobbin. The bobbin is rotated in a double-axis gyratory motion (a cardioid), which causes a variable gravity (G) field to act on the column during each rotation. This motion causes the column to see one partitioning step per revolution and components of the sample separate in the column due to their partitioning coefficient between the two immiscible liquid phases used. There are many types of CCC available today. These include HSCCC (High Speed CCC) and HPCCC (High Performance CCC). HPCCC is the latest and best performing version of the instrumentation available currently.

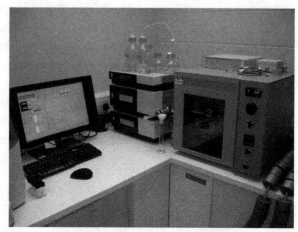

An example of a HPCCC system

Chiral Chromatography

Chiral chromatography involves the separation of stereoisomers. In the case of enantiomers, these have no chemical or physical differences apart from being three-dimensional mirror images. Conventional chromatography or other separation processes are incapable of separating them. To enable chiral separations to take place, either the mobile phase or the stationary phase must themselves be made chiral, giving differing affinities between the analytes. Chiral chromatography HPLC columns (with a chiral stationary phase) in both normal and reversed phase are commercially available.

Affinity Chromatography

Affinity chromatography was introduced almost 50 years back as a powerful tool for purification of biologically active molecules like proteins. This technique has revolutionary impact on modern biological sciences such as molecular biology, biochemistry, medicine and biotechnology. This technique exploits molecular recognition principle of a biological compound to be separated by the specific ligand to purify it from a mixture of compounds.

The affinity chromatography is a type of liquid chromatography for the separation and specific analysis of sample components. This type of chromatography makes use of a reversible "biological interaction" (molecular recognition) for the separation and analysis of specific analytes within a sample e.g. enzyme with an inhibitor and antigen with an antibody. One of the components, the ligand is immobilized onto a solid matrix, which is then used to selectively purify the target protein. Including a competing ligand in mobile phase or changing pH that elutes the target protein out. For example, Ni-Affinity chromatography is applied for the purification of 6xHis tagged proteins in which Ni is the chelating metal which is attached on NTA matrix.

Theoretically affinity chromatography is capable of giving absolute purification in a single step. The technique was developed for purification of enzymes but now affinity chromatography is used for various other purposes like purification of nucleotides, nucleic acid, immunoglobulin, membrane receptors etc.

Affinity chromatography is a method of separating biochemical mixtures based on a highly specific interaction between antigen and antibody, enzyme and substrate, or receptor and ligand. It is a type of chromatographic lab technique used for purifying biological molecules within a mixture by exploiting molecular properties. Biological macromolecules such as enzymes and other proteins, interact with other molecules with high specificity through several different types of bonds and interaction. Such interactions including hydrogen bonding, ionic interaction, disulfide bridges, hydrophobic interaction, and more. The high selectivity of affinity chromatography is caused by allowing the desired molecule to interact with the stationary phase and be bound within the column in order to be separated from the undesired material which will not interact and elute first.

Uses

Affinity chromatography can be used to:

- Purify and concentrate a substance from a mixture into a buffering solution
- Reduce the amount of a the unwanted substances in a mixture
- Discern what biological compounds bind to a particular substance
- Purify and concentrate an enzyme solution.

Principle

The stationary phase is typically a gel matrix, often of agarose; a linear sugar molecule derived from algae. Usually the starting point is an undefined heterogeneous group of molecules in solution, such as a cell lysate, growth medium or blood serum. The molecule of interest will have a well known and defined property, and can be exploited during the affinity purification process. The process itself can be thought of as an entrapment, with the target molecule becoming trapped on a solid or stationary phase or medium. The other molecules in the mobile phase will not become trapped as they do not possess this property. The stationary phase can then be removed from the mixture, washed and the target molecule released from the entrapment in a process known as dialysis. Possibly the most common use of affinity chromatography is for the purification of recombinant proteins.

Batch and Column Setups

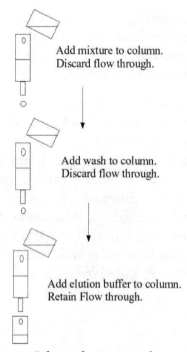

Column chromatography

Binding to the solid phase may be achieved by column chromatography whereby the solid medium is packed onto a column, the initial mixture run through the column to allow setting, a wash buffer run through the column and the elution buffer subsequently applied to the column and collected. These steps are usually done at ambient pressure. Alternatively, binding may be achieved using a batch treatment, for example, by adding the initial mixture to the solid phase in a vessel, mixing, separating the solid phase, removing the liquid phase, washing, re-centrifuging, adding the elution buffer, re-centrifuging and removing the elute.

Sometimes a hybrid method is employed such that the binding is done by the batch method, but the solid phase with the target molecule bound is packed onto a column and washing and elution are done on the column.

A third method, expanded bed absorption, which combines the advantages of the two methods mentioned above, has also been developed. The solid phase particles are placed in a column where liquid phase is pumped in from the bottom and exits at the top. The gravity of the particles ensure that the solid phase does not exit the column with the liquid phase.

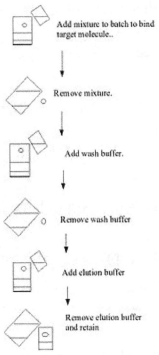

Batch chromatography

Affinity columns can be eluted by changing salt concentrations, pH, pI, charge and ionic strength directly or through a gradient to resolve the particles of interest.

More recently, setups employing more than one column in series have been developed. The advantage compared to single column setups is that the resin material can be fully loaded, since non-binding product is directly passed on to a consecutive column with fresh column material. The resin costs per amount of produced product can thus be drastically reduced. Since one column can always be eluted and regenerated while the other column is loaded, already two columns are sufficient to make full use of the advantages. Additional columns can give additional flexibility for elution and regeneration times, at the cost of additional equipment and resin costs.

Specific Uses

Affinity chromatography can be used in a number of applications, including nucleic acid purification, protein purification from cell free extracts, and purification from blood.

Various Affinity Media

Many different affinity media exist for a variety of possible uses. Briefly, they are (generalized):

- Activated/Functionalized – Works as a functional spacer, support matrix, and eliminates handling of toxic reagents.

- Amino Acid – Used with a variety of serum proteins, proteins, peptides, and enzymes, as well as rRNA and dsDNA.

- Avidin Biotin – Used in the purification process of biotin/avidin and their derivatives.

- Carbohydrate Bonding – Most often used with glycoproteins or any other carbohydrate-containing substance.

- Carbohydrate – Used with lectins, glycoproteins, or any other carbohydrate metabolite protein.

- Dye Ligand – This media is nonspecific, but mimics biological substrates and proteins.

- Glutathione – Useful for separation of GST tagged recombinant proteins.

- Heparin – This media is a generalized affinity ligand, and it is most useful for separation of plasma coagulation proteins, along with nucleic acid enzymes and lipases.

- Hydrophobic Interaction – Most commonly used to target free carboxyl groups and proteins.

- Immunoaffinity – Detailed below, this method utilizes antigens' and antibodies' high specificity to separate.

- Immobilized Metal Affinity Chromatography – Detailed further below, this method uses interactions between metal ions and proteins (usually specially tagged) to separate.

- Nucleotide/Coenzyme – Works to separate dehydrogenases, kinases, and transaminases.

- Nucleic Acid – Functions to trap mRNA, DNA, rRNA, and other nucleic acids/oligonucleotides.

- Protein A/G – This method is used to purify immunoglobulins.

- Speciality – Designed for a specific class or type of protein/coenzyme, this type of media will only work to separate a specific protein or coenzyme.

Immunoaffinity

Another use for the procedure is the affinity purification of antibodies from blood serum. If serum is known to contain antibodies against a specific antigen (for example if the serum comes from an organism immunized against the antigen concerned) then it can be used for the affinity purification of that antigen. This is also known as Immunoaffinity Chromatography. For example, if an organism is immunised against a GST-fusion protein it will produce antibodies against the fusion-protein, and possibly antibodies against the GST tag as well. The protein can then be covalently coupled to a solid support such as agarose and used as an affinity ligand in purifications of antibody from immune serum.

For thoroughness the GST protein and the GST-fusion protein can each be coupled separately. The serum is initially allowed to bind to the GST affinity matrix. This will remove antibodies against the GST part of the fusion protein. The serum is then separated from the solid support and allowed to bind to the GST-fusion protein matrix. This allows any antibodies that recognize the antigen to be captured on the solid support. Elution of the antibodies of interest is most often achieved using a low pH buffer such as glycine pH 2.8. The eluate is collected into a neutral tris or phosphate buffer, to neutralize the low pH elution buffer and halt any degradation of the antibody's activity. This is a nice example as affinity purification is used to purify the initial GST-fusion protein, to remove the undesirable anti-GST antibodies from the serum and to purify the target antibody.

A simplified strategy is often employed to purify antibodies generated against peptide antigens. When the peptide antigens are produced synthetically, a terminal cysteine residue is added at either the N- or C-terminus of the peptide. This cysteine residue contains a sulfhydryl functional group which allows the peptide to be easily conjugated to a carrier protein (e.g. Keyhole limpet hemocyanin (KLH)). The same cysteine-containing peptide is also immobilized onto an agarose resin through the cysteine residue and is then used to purify the antibody.

Most monoclonal antibodies have been purified using affinity chromatography based on immunoglobulin-specific Protein A or Protein G, derived from bacteria.

Immobilized Metal Ion Affinity Chromatography

A chromatography column containing nickel-agarose beads used for purification of proteins with histidine tags

Immobilized metal ion affinity chromatography (IMAC) is based on the specific coordinate covalent bond of amino acids, particularly histidine, to metals. This technique works by allowing proteins with an affinity for metal ions to be retained in a column containing immobilized metal ions, such as cobalt, nickel, copper for the purification of histidine containing proteins or peptides, iron, zinc or gallium for the purification of phosphorylated proteins or peptides. Many naturally occurring proteins do not have an affinity for metal ions, therefore recombinant DNA technology can be used to introduce such a protein tag into the relevant gene. Methods used to elute the protein of interest include changing the pH, or adding a competitive molecule, such as imidazole.

Recombinant Proteins

Possibly the most common use of affinity chromatography is for the purification of recombinant proteins. Proteins with a known affinity are protein tagged in order to aid their purification. The protein may have been genetically modified so as to allow it to be selected for affinity binding; this is known as a fusion protein. Tags include glutathione-S-transferase (GST), hexahistidine (His), and maltose binding protein (MBP). Histidine tags have an affinity for nickel or cobalt ions which have been immobilized by forming coordinate covalent bonds with a chelator incorporated in the stationary phase. For elution, an excess amount of a compound able to act as a metal ion ligand, such as imidazole, is used. GST has an affinity for glutathione which is commercially available immobilized as glutathione agarose. During elution, excess glutathione is used to displace the tagged protein.

Lectins

Lectin affinity chromatography is a form of affinity chromatography where lectins are used to separate components within the sample. Lectins, such as Concanavalin A are proteins which can bind specific carbohydrate (sugar) molecules. The most common application is to separate glycoproteins from non-glycosylated proteins, or one glycoform from another glycoform.

Specialty

Another use for affinity chromatography is the purification of specific proteins using a gel matrix that is unique to a specific protein. For example, the purification of E.Coli-B-Galactosidase is accomplished by affinity chromatography using P-Aminobenyl-1-Thio-B-D-Galactopyranosyl Agarose as the affinity matrix. P-Aminobenyl-1-Thio-B-D-Galactopyranosyl Agarose is used as the affinity matrix because it contains a galactopyranosyl group, which serves as a good substrate analog for E.Coli-B-Galactosidase. This property allows the enzyme to bind to the stationary phase of the affinity matrix and is eluted by adding increasing concentrations of salt to the column.

The biological interactions involve mostly non covalent interactions between the reactive groups of molecule targeted for purification and ligand with a dissociation constant K_d.

$$K_d = \frac{[A][B]}{[AB]}$$

Where, A is assumed as molecule targeted and B as ligand and AB is the complex formed between them. K_d varies between 10^{-3} to 10^{-7} M for affinity binding.

The principle of affinity chromatography is that the stationary phase consists of a support medium (e.g. cellulose beads) on which the substrate (or sometimes a coenzyme) has been bound covalently, in such a way that the reactive groups that are essential for enzyme binding are exposed. As the crude mixture of proteins is passed through the chromatography column, proteins with binding site for the immobilized substrate will bind to the stationary phase, while all other proteins will be eluted in the void volume of the column.

Once the other proteins have all been eluted, the bound enzyme(s) can be eluted in various ways.

Procedure

Immobilizing protein ligand on solid support

Preparation of column: The solid support (bead matrix) is a gel loaded into an elution column. Sepharose, agarose and cellulose are the most commonly used solid support, because the hydroxyl groups on the sugar residues can be easily manipulated to accept a ligand. The ligand is then selected according to the desired isolate. In case a researcher planning to isolate antibodies specific for antigen A from an antiserum, antigen A can be used as ligand. There are several methods of immobilizing protein ligand on these solid support and some examples are given below.

Sometimes flexible spacer - arm is attached between ligand and solid support to render better flexibility to ligand. For example NHS-activated Sepharose (agarose beads with 10-atom spacer arms (6-aminohexanoic acid) attached by epichlorohydrin and activated by N-hydroxysuccinimide) is used for protein ligand immobilization.

Flexible spacer - arm may be attached between ligand and solid support

Likewise, if someone wants to separate specific enzyme, a strategy may involve immobilization of either its substrate, an inhibitor, or even a cofactor on solid support.

Two factors are required for the ligand:

I. Specific and reversible binding with the desired protein.

II. The ligand is capable of covalent bonding to the matrix without disrupting its binding activity. This is usually facilitated by the placement of spacer arms between the ligand and the matrix, so that in case the active site is buried deep within the ligand, it is not physically hidden from its binding substrate.

Steps of Affinity Purification

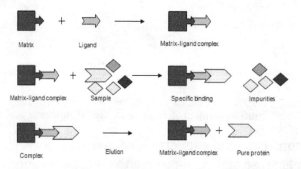

An outline of affinity purification scheme.

Loading of solution containing the substance to be isolated: The solution is usually a protein rich mixture such as antiserum, which is poured into the elution column and allowed to run through the gel, at a controlled rate. Proteins with specific affinity for immobilized ligand shall bind and other proteins will go in flow through. This follows washing of column with buffer to remove all unbound protein. Some affinity purification procedures are summarized in the table.

Table: List of a few affinity procedures.

Ligand	Affinity/purification process
Avidin	Avidin is a tetrameric protein deposited in the egg white of birds, reptile and amphibians. This protein has affinity for biotin, cofactor of several enzymes. Immobilized avidin column is used for purification of biotin containing enzyme. N on-biotin molecules do not bind to immobilized avidin and are washed away. Bound proteins with biotin may be competitively eluted using 2 mM biotin.
Calmodulin	Calmodulin, a regulatory Ca^{+2} binding protein, is present in all eukaryotic cells. Calmodulin binds proteins through their interactions with hydrophobic sites on its surface which is exposed after Ca^{+2} binding to the enzyme. Elustion is done by chelating agent such as EGTA or EDTA. Once Ca^{+2} bound to Calmodulin is chelated, a reversal of the conformational change which expose the protein binding sites takes place. This results in protein elution.
Concanavalin A	Concanavalin A is tetrameric metalloprotein. This protein is a carbohydrate-binding protein (lectin) originally lycoproteins containing a-D-mannopyanosyl and a-D-glucopyranosyl residues. Elusion of bound glycoprotein is achieved by increasing gradient of α-D- methylmannoside or α -D- methylglucoside.
Protein A and G	Immunoglobulins
Cibacron Blue F3G-A	Cibacron Blue F3G-A is a sulfonated triazine dye that can be immobilized on a solid support and used for affinity chromatography of nucleotide-requiring enzymes.

Affinity Chromatography-II

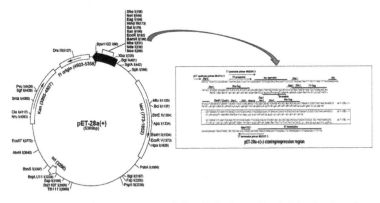

pET 28(a) expression vector and detailed view of multiple cloning sites

Because of advancement in Molecular Biology now it is possible to identify gene of a given protein. It is also possible to transfer the gene to other organism (for example E. coli) in a vector (simply a vehicle to transfer foreign gene into organism or another cell) and express in the same organism. Such proteins are called recombinant protein.

Genetic manipulations are not scope of this course but students might have studied this in Molecular Biology course. Since purification of a protein can be a complex and time-consuming process so that the expression vectors are designed for higher level of expression of recombinant proteins with tags to facilitate the further purification. The DNA sequence codes for the protein are cloned in expression vectors at multiple cloning sites in continuation with tags either at N-terminal or C-terminal. These tags are also DNA sequences which code a small peptide or even a small protein to facilitate the purification of the recombinant protein. e.g. 6x His Tag or GST Tag are commonly used for purifying recombinant proteins.

His-Tag for Purification of Recombinant Proteins: A hexa-His sequence is called a His-Tag. It has been shown that an amino acid sequence consisting of 6 or more His residues in a row will also act as a metal binding site for a recombinant protein. The 6xHis affinity tag facilitates binding to immobilized Ni. 6xHis coding sequence can be placed at the C- or N-terminus of the protein of interest and recombinant protein with 6xHis residues can be expressed. The tag is poorly immunogenic, and at pH 8.0 the tag is small, uncharged, and therefore does not generally affect secretion, compartmentalization, or folding of the fusion protein within the cell. In most cases, the 6xHis tag does not interfere with the structure or function of the purified protein as demonstrated for a wide variety of proteins, including enzymes, transcription factors, and vaccines. Sometime, a protease cleavage site is inserted between protein sequence and His-tag. After His-tag affinity purification, the purified protein is treated with the specific protease to remove His-tag. A very common example is His -tagged Proteins with Thrombin, a protease, cleavage site.

Thrombin recognizes the consensus sequence Leu-Val-Pro-Arg-Gly-Ser, cleaving the peptide bond between Arg and Gly. This is utilised in many vector systems which encode such a protease cleavage site allowing removal of an upstream domain.

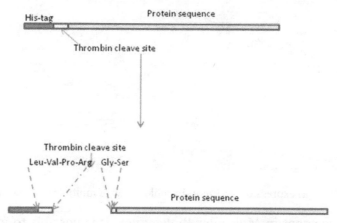

His - tagged Proteins with Thrombin cleavage site. First, his-tagged protein is purified by Immobilized-metal affinity chromatography. After purification, tag is removed by thrombin cleavage. After thrombin cleavage, the sample is again passed through Immobilized-metal affinity chromatography column. The cleaved protein will come in elution while the tag will remain bound to column

Immobilized-metal affinity chromatography (IMAC) was first used to purify proteins in 1975 by Porath and group using the chelating ligand iminodiacetic acid (IDA, Figure). IDA was charged with metal ions such as Ni^{2+}, and then used to purify a variety of different proteins and peptides. IDA has only 3 metal-chelating sites and cannot tightly bind metal ions. Weak binding leads to ion leaching upon loading with strongly chelating proteins and peptides or during wash steps. This results in low yields, and metal-ion contamination of isolated proteins.

Comparison of interaction of different metal chelate matrices with nickel ions

Nitrilotriacetic acid (NTA), is a tetradentate chelating adsorbent developed by Hoffmann-La Roche that overcomes these problems. NTA occupies four of the six ligand binding sites in the coordination sphere of the nickel ion, leaving two sites free to interact with the 6xHis tag. NTA binds metal ions far more strongly than other available chelating resins and retains the ions under a wide variety of conditions, especially under stringent wash conditions. The unique, patented NTA matrices can therefore bind 6xHis-tagged proteins more tightly than IDA matrices, allowing the purification of proteins from less than 1% of the total protein preparation to more than 95% homogeneity in just one step.

Binding of His-tag with Ni-NTA groups

Protein binding: Proteins containing one or more 6xHis affinity tags, located at either the amino and/or carboxyl terminus of the protein, can bind to the Ni-NTA groups on the matrix with an affinity far greater than that of antibody–antigen or enzyme–substrate interactions. Binding of the 6xHis tag does not depend on the three-dimensional structure of the protein. Even when the tag is not completely accessible it will bind as long as more than two histidine residues are available to interact with the nickel ion; in general, the smaller the number of accessible histidine residues, the weaker the binding will be. Untagged proteins that have histidine residues in close proximity on their surface may also bind to Ni-NTA, but in most cases this interaction will be much weaker than the binding of the 6xHis tag. Any host proteins that bind nonspecifically to the NTA resin itself can be easily washed away under relatively stringent conditions that do not affect the binding of 6xHis-tagged proteins. Binding can be carried out in a batch or column mode. If the concentration of 6xHis-tagged proteins is low, or if they are expressed at low levels, or secreted into the media, the proteins should be bound to Ni-NTA in a batch procedure, and under conditions in which background proteins do not compete for the binding sites, i.e. at a slightly reduced pH or in the presence of low imidazole concentrations (10–20 mM). At low expression levels under native conditions, binding can be optimized for every protein by adjusting the imidazole concentration and/or pH of the lysis buffer. If high levels of background proteins are still present, equilibrating the Ni-NTA matrix with lysis buffer containing 10–20 mM imidazole prior to binding is recommended. The matrix is thus "shielded", and non-specific binding of proteins that weakly interact is significantly reduced.

Supplementary section: Proteins may be purified on Ni-NTA resins in either a batch or a column procedure. The batch procedure entails binding the protein to the Ni-NTA resin in solution and then packing the protein–resin complex into a column for the washing and elution steps. This strategy promotes efficient binding of the 6xHis-tagged protein especially when the 6xHis tag is not fully accessible or when the protein in the lysate is present at a very low concentration. In the column procedure, the Ni-NTA resin is first packed into the column and equilibrated with the lysis buffer. The cell lysate is then slowly applied to the column. Washing and elution steps are identical in the batch and column procedure.

Wash: Endogenous proteins with histidine residues that interact with the Ni-NTA groups can be washed out of the matrix with stringent conditions achieved by adding imidazole at a 10–50 mM concentration.

Protein elution: The histidine residues in the 6xHis tag have a pKa of approximately 6.0 and will become protonated if the pH is reduced (pH 4.5–5.3). Under these conditions the 6xHis-tagged protein can no longer bind to the nickel ions and will dissociate from the Ni-NTA resin. Similarly, if the imidazole concentration is increased to 100–250 mM, the 6xHis-tagged proteins will also dissociate because they can no longer compete for binding sites on the Ni-NTA resin. Elution conditions are highly reproducible, but must be determined for each 6xHis-tagged protein being purified. Reagents such as

EDTA or EGTA chelate the nickel ions and remove them from the NTA groups. This causes the 6xHis-tagged protein to elute as a protein–metal complex. NTA resins that have lost their nickel ions become white in color and must be recharged if they are to be reused. Whereas all elution methods (imidazole, pH, and EDTA) are equally effective, imidazole is mildest and is recommended under native conditions, when the protein would be damaged by a reduction in pH, or when the presence of metal ions in the eluate may have an adverse effect on the purified protein.

Supplementary section: In bacterial expression systems, the recombinant proteins are usually expressed at high levels, and the level of copurifying contaminant proteins is relatively low. Therefore it is generally not necessary to wash the bound 6xHis-tagged protein under very stringent conditions. In lysates derived from eukaryotic expression systems the relative abundance of proteins that may contain neighbouring histidines is higher; the resulting background problem becomes more critical especially when non-denaturing procedures are employed. In these instances it becomes necessary to use imidazole gradient.

Additional points: As low affinity binding with other contaminant protein is also possible, low concentrations of imidazole in the lysis and wash buffers (10–20 mM) are recommended. The imidazole ring is part of the structure of histidine. The imidazole rings in the histidine residues of the 6xHis tag bind to the nickel ions immobilized by the NTA groups on the matrix. Imidazole itself can also bind to the nickel ions and disrupt the binding of dispersed histidine residues in nontagged background proteins. At low imidazole concentrations, nonspecific, low affinity binding of background proteins is prevented, while 6xHis-tagged proteins still bind strongly to the Ni-NTA matrix. Therefore, adding imidazole to the lysis buffer leads to greater purity in fewer steps. For most proteins, up to 20 mM imidazole can be used without affecting the yield. If the tagged protein does not bind under these conditions, the amount of imidazole should be reduced to 1–5 mM.

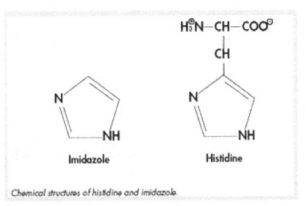

Chemical structures of histidine and imidazole.

Binding of tagged proteins to Ni-NTA resin is not conformation-dependent and is not affected by most detergents and denaturants. The stability of the 6xHis–Ni-NTA interaction in the presence of low levels of β-ME (up to 20 mM) in the lysis buffer can be used to prevent the copurification of host proteins that may have formed disulfide bonds with the protein of interest during cell lysis. Detergents such as Triton X-100 and Tween 20 (up to 2%), or high salt concentrations (up to 2 M NaCl), also have no effect on binding, and may reduce nonspecific binding to the matrix due to nonspecific hydrophobic or ionic interactions. Nucleic acids that might associate with certain DNA and RNA-binding proteins are also removed without effecting the recovery of the 6xHis-tagged protein.

Glutathione S-transferase

Glutathione S-transferases (GSTs), previously known as ligandins, comprise a family of eukaryotic and prokaryotic phase II metabolic isozymes best known for their ability to catalyze the conjugation of the reduced form of glutathione (GSH) to xenobiotic substrates for the purpose of detoxification. The GST family consists of three superfamilies: the cytosolic, mitochondrial, and microsomal—also known as MAPEG—proteins. Members of the GST superfamily are extremely diverse in amino acid sequence, and a large fraction of the sequences deposited in public databases are of unknown function. The Enzyme Function Initiative (EFI) is using GSTs as a model superfamily to identify new GST functions.

GSTs can constitute up to 10% of cytosolic protein in some mammalian organs. GSTs catalyse the conjugation of GSH — via a sulfhydryl group — to electrophilic centers on a wide variety of substrates in order to make the compounds more water-soluble. This activity detoxifies endogenous compounds such as peroxidised lipids and enables the breakdown of xenobiotics. GSTs may also bind toxins and function as transport proteins, which gave rise to the early term for GSTs, *ligandin*.

Classification

Protein sequence and structure are important additional classification criteria for the three superfamilies (cytosolic, mitochondrial, and MAPEG) of GSTs: while classes from the cytosolic superfamily of GSTs possess more than 40% sequence homology, those from other classes may have less than 25%. Cytosolic GSTs are divided into 13 classes based upon their structure: alpha, beta, delta, epsilon, zeta, theta, mu, nu, pi, sigma, tau, phi, and omega. Mitochondrial GSTs are in class kappa. The MAPEG superfamily of microsomal GSTs consists of subgroups designated I-IV, between which amino acid sequences share less than 20% identity. Human cytosolic GSTs belong to the alpha, zeta, theta, mu, pi, sigma, and omega classes, while six isozymes belonging to classes I, II, and IV of the MAPEG superfamily are known to exist.

Nomenclature

Standardized GST nomenclature first proposed in 1992 identifies the species to which the isozyme of interest belongs with a lower-case initial (e.g., "h" for human), which precedes the abbreviation GST. The isozyme class is subsequently identified with an upper-case letter (e.g., "A" for alpha), followed by an Arabic numeral representing the class subfamily (or subunit). Because both mitochondrial and cytosolic GSTs exist as dimers, and only heterodimers form between members of the same class, the second subfamily component of the enzyme dimer is denoted with a hyphen, followed by an additional Arabic numeral. Therefore, if a human glutathione *S*-transferase is a homodimer in the pi-class subfamily 1, its name will be written as "hGSTP1-1."

The early nomenclature for GSTs referred to them as "Y" proteins, referring to their separation in the "Y" fraction (as opposed to the "X and Z" fractions) using Sephadex G75 chromatography. As GST sub-units were identified they were referred to as Ya, Yp, etc. with if necessary, a number identifying the monomer isoform (e.g. Yb1). Litwack *et al* proposed the term "Ligandin" to cover the proteins previously known as "Y" proteins.

In clinical chemistry and toxicology, the terms alpha GST, mu GST, and pi GST are most commonly used.

Structure

The glutathione binding site, or "G-site," is located in the thioredoxin-like domain of both cytosolic and mitochondrial GSTs. The region containing the greatest amount of variability between the assorted classes is that of helix α2, where one of three different amino acid residues interacts with the glycine residue of glutathione. Two subgroups of cytosolic GSTs have been characterized based upon their interaction with glutathione: the Y-GST group, which uses a tyrosine residue to activate glutathione, and the S/C-GST, which instead uses serine or cysteine residues.

"GST proteins are globular proteins with an N-terminal mixed helical and beta-strand domain and an all-helical C-terminal domain."

The porcine pi-class enzyme pGTSP1-1 was the first GST to have its structure determined, and it is representative of other members of the cytosolic GST superfamily, which contain a thioredoxin-like N-terminal domain as well as a C-terminal domain consisting of alpha helices.

Mammalian cytosolic GSTs are dimeric, with both subunits being from the same class of GSTs, although not necessarily identical. The monomers are approximately 25 kDa in size. They are active over a wide variety of substrates with considerable overlap. The following table lists all GST enzymes of each class known to exist in *Homo sapiens*, as found in the UniProtKB/Swiss-Prot database.

GST Class	*Homo sapiens* GST Class Members (22)
Alpha	GSTA1, GSTA2, GSTA3, GSTA4, GSTA5
Delta	
Kappa	GSTK1

Mu	GSTM1, GSTM1L (RNAi), GSTM2, GSTM3, GSTM4, GSTM5
Omega	GSTO1, GSTO2
Pi	GSTP1
Theta	GSTT1, GSTT2, GSTT4
Zeta	GSTZ1 (aka GSTZ1 MAAI-Maleylacetoacetate isomerase)
Microsomal	MGST1, MGST2, MGST3

Function

The activity of GSTs is dependent upon a steady supply of GSH from the synthetic enzymes gamma-glutamylcysteine synthetase and glutathione synthetase, as well as the action of specific transporters to remove conjugates of GSH from the cell. The primary role of GSTs is to detoxify xenobiotics by catalyzing the nucleophilic attack by GSH on electrophilic carbon, sulfur, or nitrogen atoms of said nonpolar xenobiotic substrates, thereby preventing their interaction with crucial cellular proteins and nucleic acids. Specifically, the function of GSTs in this role is twofold: to bind both the substrate at the enzyme's hydrophobic H-site and GSH at the adjacent, hydrophilic G-site, which together form the active site of the enzyme; and subsequently to activate the thiol group of GSH, enabling the nucleophilic attack upon the substrate. The glutathione molecule binds in a cleft between N and C-terminal domains - the catalytically important residues are proposed to reside in the N-terminal domain. Both subunits of the GST dimer, whether hetero- or homodimeric in nature, contain a single nonsubstrate binding site, as well as a GSH-binding site. In heterodimeric GST complexes such as those formed by the cytosolic mu and alpha classes, however, the cleft between the two subunits is home to an additional high-affinity nonsubstrate xenobiotic binding site, which may account for the enzymes' ability to form heterodimers.

The compounds targeted in this manner by GSTs encompass a diverse range of environmental or otherwise exogenous toxins, including chemotherapeutic agents and other drugs, pesticides, herbicides, carcinogens, and variably-derived epoxides; indeed, GSTs are responsible for the conjugation of β_1-8,9-epoxide, a reactive intermediate formed from aflatoxin B_1, which is a crucial means of protection against the toxin in rodents. The detoxification reactions comprise the first four steps of mercapturic acid synthesis, with the conjugation to GSH serving to make the substrates more soluble and allowing them to be removed from the cell by transporters such as multidrug resistance-associated protein 1 (MRP1). After export, the conjugation products are converted into mercapturic acids and excreted via the urine or bile.

Most mammalian isoenzymes have affinity for the substrate 1-chloro-2,4-dinitrobenzene, and spectrophotometric assays utilising this substrate are commonly used to report GST activity. However, some endogenous compounds, e.g., bilirubin, can inhibit the activity of GSTs. In mammals, GST isoforms have cell specific distributions (e.g., alpha GST in hepatocytes and pi GST in the biliary tract of the human liver).

Role in Cell Signaling

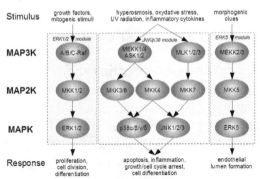

A simplified overview of MAPK pathways in mammals, organised into three main signaling modules (ERK1/2, JNK/p38 and ERK5)

Although best known for their ability to conjugate xenobiotics to GSH and thereby detoxify cellular environments, GSTs are also capable of binding nonsubstrate ligands, with important cell signaling implications. Several GST isozymes from various classes have been shown to inhibit the function of a kinase involved in the MAPK pathway that regulates cell proliferation and death, preventing the kinase from carrying out its role in facilitating the signaling cascade.

Cytosolic GSTP1-1, a well-characterized isozyme of the mammalian GST family, is expressed primarily in heart, lung, and brain tissues; in fact, it is the most common GST expressed outside the liver. Based on its overexpression in a majority of human tumor cell lines and prevalence in chemotherapeutic-resistant tumors, GSTP1-1 is thought to play a role in the development of cancer and its potential resistance to drug treatment. Further evidence for this comes from the knowledge that GSTP can selectively inhibit C-jun phosphorylation by JNK, preventing apoptosis. During times of low cellular stress, a complex forms through direct protein–protein interactions between GSTP and the C-terminus of JNK, effectively preventing the action of JNK and thus its induction of the JNK pathway. Cellular oxidative stress causes the dissociation of the complex, oligomerization of GSTP, and induction of the JNK pathway, resulting in apoptosis. The connection between GSTP inhibition of the pro-apoptotic JNK pathway and the isozyme's overexpression in drug-resistant tumor cells may itself account for the tumor cells' ability to escape apoptosis mediated by drugs that are not substrates of GSTP.

Like GSTP, GSTM1 is involved in regulating apoptotic pathways through direct protein–protein interactions, although it acts on ASK1, which is upstream of JNK. The mechanism and result are similar to that of GSTP and JNK, in that GSTM1 sequesters ASK1 through complex formation and prevents its induction of the pro-apoptotic p38 and JNK portions of the MAPK signaling cascade. Like GSTP, GSTM1 interacts with its partner in the absence of oxidative stress, although ASK1 is also involved in heat shock response, which is likewise prevented during ASK1 sequestration. The fact that high levels of GST are associated with resistance to apoptosis induced by a range of substances, including chemotherapeutic agents, supports its putative role in MAPK signaling prevention.

Implications in Cancer Development

There is a growing body of evidence supporting the role of GST, particularly GSTP, in cancer development and chemotherapeutic resistance. The link between GSTP and cancer is most obvious in the overexpression of GSTP in many cancers, but it is also supported by the fact that the transformed phenotype of tumor cells is associated with aberrantly regulated kinase signaling pathways and cellular addiction to overexpressed proteins. That most anti-cancer drugs are poor substrates for GSTP indicates that the role of elevated GSTP in many tumor cell lines is not to detoxify the compounds, but must have another purpose; this hypothesis is also given credence by the common finding of GSTP overexpression in tumor cell lines that are not drug resistant.

Clinical Significance

In addition to their roles in cancer development and chemotherapeutic drug resistance, GSTs are implicated in a variety of diseases by virtue of their involvement with GSH. Although the evidence is minimal for the influence of GST polymorphisms of the alpha, mu, pi, and theta classes on susceptibility to various types of cancer, numerous studies have implicated such genotypic variations in asthma, atherosclerosis, allergies, and other inflammatory diseases.

Because diabetes is a disease that involves oxidative damage, and GSH metabolism is dysfunctional in diabetic patients, GSTs may represent a potential target for diabetic drug treatment. In addition, insulin administration is known to result in increased GST gene expression through the PI3K/AKT/mTOR pathway and reduced intracellular oxidative stress, while glucagon decreases such gene expression.

Omega-class GST (GSTO) genes, in particular, are associated with neurological diseases such as Alzheimer's, Parkinson's, and amyotrophic lateral sclerosis; again, oxidative stress is believed to be the culprit, with decreased GSTO gene expression resulting in a lowered age of onset for the diseases.

Release of GSTs as an Indication of Organ Damage

The high intracellular concentrations of GSTs coupled with their cell-specific cellular distribution allows them to function as biomarkers for localising and monitoring injury to defined cell types. For example, hepatocytes contain high levels of alpha GST and serum alpha GST has been found to be an indicator of hepatocyte injury in transplantation, toxicity and viral infections.

Similarly, in humans, renal proximal tubular cells contain high concentrations of alpha

GST, while distal tubular cells contain pi GST. This specific distribution enables the measurement of urinary GSTs to be used to quantify and localise renal tubular injury in transplantation, nephrotoxicity and ischaemic injury.

In rodent pre-clinical studies, urinary and serum alpha GST have been shown to be sensitive and specific indicators of renal proximal tubular and hepatocyte necrosis respectively.

GST-tags and the GST pull-down Assay

GST can be added to a protein of interest to purify it from solution in a process known as a pull-down assay. This is accomplished by inserting the GST DNA coding sequence next to that which codes for the protein of interest. Thus, after transcription and translation, the GST protein and the protein of interest will be expressed together as a fusion protein. Because the GST protein has a strong binding affinity for GSH, beads coated with the compound can be added to the protein mixture; as a result, the protein of interest attached to the GST will stick to the beads, isolating the protein from the rest of those in solution. The beads are recovered and washed with free GST to detach the protein of interest from the beads, resulting in a purified protein. This technique can be used to elucidate direct protein–protein interactions. A drawback of this assay is that the protein of interest is attached to GST, altering its native state.

A GST-tag is often used to separate and purify proteins that contain the GST-fusion protein. The tag is 220 amino acids (roughly 26 KDa) in size, which, compared to tags such as the Myc-tag or the FLAG-tag, is quite large. It can be fused to either the N-terminus or C-terminus of a protein. However, many commercially available sources of GST-tagged plasmids include a thrombin domain for cleavage of the GST tag during protein purification.

Affinity Chromatography-III

Glutathione S-transferase for purification of recombinant proteins: For easy purification of recombinant proteins they are generally tagged with GST (Glutathione S-transferase) protein to create fusion proteins (protein sequence attached with GST sequence). The tag has the size of 220 amino acids (roughly 26 kDa), which is quite big compared to other tags like the myc- or FLAG-tag. It generally helps the recombinant protein for soluble expression and further purification. Moreover, a thrombin (a protease) recognition sequence is included in between the GST tag and protein sequence. This helps in removal of GST tag by cleavage with thrombin after purification of fusion protein.

The GST part of fusion proteins has affinity for glutathione as glutathione is the substrate for GST. This enzyme (GST) substrate (glutathione) affinity is used for purification of fusion protein. Agarose or other polymer beads can be coated with glutathione, and such glutathione-agarose beads bind GST-proteins. The crude cell lysate can be loaded on a column packed with glutathione-agarose matrix, and washed extensively. The fusion protein will bind with glutathione coated agarose beads through GST, while other proteins will wash off. The elution of bind fusion protein was performed by free glutathione solution. Due to higher concentration of glutathione in solution, fusion protein leaves the glutathione coated beads and comes in solution.

The eluted affinity purified fusion protein will now be subjected to thrombin cleavage for removal of GST tag. Amount of thrombin used for cleave is very minute and generally removal of thrombin is not required (or an immobilized thrombin may be used which may be removed by simple centrifugation after completion of cleavage reaction). The thrombin treated protein was again loaded on regenerated column. This time the cleaved GST tag will bind with beads and recombinant protein will come in un-bound fraction. A general scheme for purification of proteins with affinity tag is given is the Figure.

Maltose-Binding Protein (MBP) for purification of recombinant proteins: In this technique the gene of interest is cloned into pMAL vector creating MBP-encoding malE gene and factor Xa (a protease) cleavage site. This gene can be expressed in E. coli producing MBP fusion protein (MBP fused with protein of interest containing factor Xa cleavage sequence between MBP and protein of interest). This MBP-fusion protein is purified using amylose column, MBP has affinity for the amylose ligand and finally fusion protein can be eluted using maltose gradient. Finally MBP tag can be cleaved from fusion protein using factor Xa protease (as there is factor Xa cleavage site between MBP and protein of interest). Generally factor Xa protease used for the cleavage is in minute amount and removal is not required (or an immobilized factor Xa protease may be used which may be removed by simple centrifugation after completion of cleavage reaction). Cleaved MBP tag can be separated from the protein of interest by loading it again to amylose column. This time cleaved MBP tag will bind to column but protein of interest will go in unbound fraction.

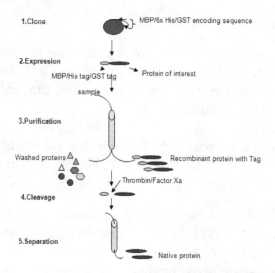

Purification of recombinant proteins using different affinity tags.

Scheme of an affinity purification method is given in following figure. This could be a very effective method to purify a recombinant protein for which there is no known easy way to purify using substrate-based affinity chromatography.

Application of Affinity Chromatography in Proteomics

Proteomics is the study of total protein pool from cell line, tissue or organism. The most commonly used experimental techniques in proteomics are 2 DE (two dimensional gel electrophoresis) for the separation of proteins from a mixture containing thousands of proteins (one dimension is isoelectric focusing and second dimension is SDS PAGE) and Mass Spectrometry for the identification of separated proteins. In the separation procedure, affinity chromatography can also play an important role.

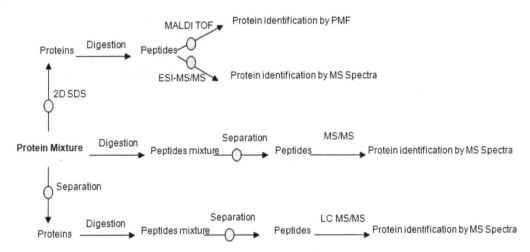

Application of affinity chromatography at different stages of proteomic analysis.
Solid circles indicate the position where affinity chromatography can be used

References

- Ninfa, Alexander J.; Ballou, David P.; Benore, Marilee (2009). Fundamental Laboratory Approaches for Biochemistry and Biotechnology (2 ed.). Wiley. p. 133. ISBN 9780470087664
- Ettre, L. S.; Sakodynskii, K. I. (March 1993). "M. S. Tswett and the discovery of chromatography II: Completion of the development of chromatography (1903–1910)". Chromatographia. 35 (5-6): 329–338. doi:10.1007/BF02277520
- Beckett GJ, Hayes JD (1987). "Glutathione S-transferase measurements and liver disease in man". Journal of Clinical Biochemistry and Nutrition. 2: 1–24. doi:10.3164/jcbn.2.1
- M., Grisham, Charles (2013-01-01). Biochemistry. Brooks/Cole, Cengage Learning. ISBN 1133106293. OCLC 777722371
- Ettre, L. S. (1993). "Nomenclature for chromatography (IUPAC Recommendations 1993)". Pure and Applied Chemistry. 65 (4). doi:10.1351/pac199365040819
- McMurry, John (2011). Organic chemistry: with biological applications (2nd ed.). Belmont, CA: Brooks/Cole. p. 395. ISBN 9780495391470

- Still, W. C.; Kahn, M.; Mitra, A. (1978). "Rapid chromatographic technique for preparative separations with moderate resolution". J. Org. Chem. 43 (14): 2923–2925. doi:10.1021/jo00408a041

- Baur, Daniel; Angarita, Monica; Müller-Späth, Thomas; Steinebach, Fabian; Morbidelli, Massimo (2016). "Comparison of batch and continuous multi-column protein A capture processes by optimal design". Biotechnology Journal. John Wiley & Sons, Inc. 11 (7): 920–931. doi:10.1002/biot.201500481. Retrieved 16 August 2016

PERMISSIONS

All chapters in this book are published with permission under the Creative Commons Attribution Share Alike License or equivalent. Every chapter published in this book has been scrutinized by our experts. Their significance has been extensively debated. The topics covered herein carry significant information for a comprehensive understanding. They may even be implemented as practical applications or may be referred to as a beginning point for further studies.

We would like to thank the editorial team for lending their expertise to make the book truly unique. They have played a crucial role in the development of this book. Without their invaluable contributions this book wouldn't have been possible. They have made vital efforts to compile up to date information on the varied aspects of this subject to make this book a valuable addition to the collection of many professionals and students.

This book was conceptualized with the vision of imparting up-to-date and integrated information in this field. To ensure the same, a matchless editorial board was set up. Every individual on the board went through rigorous rounds of assessment to prove their worth. After which they invested a large part of their time researching and compiling the most relevant data for our readers.

The editorial board has been involved in producing this book since its inception. They have spent rigorous hours researching and exploring the diverse topics which have resulted in the successful publishing of this book. They have passed on their knowledge of decades through this book. To expedite this challenging task, the publisher supported the team at every step. A small team of assistant editors was also appointed to further simplify the editing procedure and attain best results for the readers.

Apart from the editorial board, the designing team has also invested a significant amount of their time in understanding the subject and creating the most relevant covers. They scrutinized every image to scout for the most suitable representation of the subject and create an appropriate cover for the book.

The publishing team has been an ardent support to the editorial, designing and production team. Their endless efforts to recruit the best for this project, has resulted in the accomplishment of this book. They are a veteran in the field of academics and their pool of knowledge is as vast as their experience in printing. Their expertise and guidance has proved useful at every step. Their uncompromising quality standards have made this book an exceptional effort. Their encouragement from time to time has been an inspiration for everyone.

The publisher and the editorial board hope that this book will prove to be a valuable piece of knowledge for students, practitioners and scholars across the globe.

Index

A
Absolute Quantification, 84-85
Adsorption Chromatography, 29, 72-73
Amino Acid Residues, 6, 29, 45, 47, 49, 56, 169, 183-184, 217

B
Biochemistry, 10, 28, 30, 35, 45, 86, 131, 189, 196-197, 202, 223
Biodegradable Plastics, 31
Bioinformatics, 19, 28, 74
Biomarkers, 82, 161, 220
Biosynthesis, 9, 29, 39, 43, 49, 59, 172
Biosynthetic Pathways, 20, 39
Blue Native, 99

C
Catabolism, 36, 44
Catecholamines, 39
Cell Disruption, 23
Cell Signaling, 7, 15, 219
Cellular Functions, 13
Cellular Localization, 17-18
Chaperones, 11, 59-61, 68-69
Chemical Synthesis, 10, 42
Circular Dichroism, 53, 66-67
Column Chromatography, 70-72, 113, 194, 204
Computational Prediction, 55
Contaminant Removal, 25
Cyanine Dyes, 76, 106-107

D
Dark Proteome, 5
Data Interpretation, 107-108
Dispersion Forces, 60, 73
Driving Force, 58, 60
Drug Development, 1, 3, 121
Dual Polarisation Interferometry, 48, 67

E
Electric Quadrupole, 150, 189
Electrophoresis, 2, 4-5, 7, 16, 18, 24-25, 35, 74-76, 79, 83, 85, 87-96, 98-106, 124, 131-132, 142-143, 223
Energy Landscape, 56, 62
Expanded Genetic Code, 41

F
Fluorescence Spectroscopy, 66-67

G
Genome, 1, 3-4, 9, 18, 41, 74, 83, 170, 188
Gibbs-donnan Effect, 118
Gravitational Quadrupole, 152

H
High Time Resolution, 67
Human Brain, 30, 39
Human Nutrition, 40
Hydrophobic Effect, 59-60

I
Image Acquisition, 108
Incorrect Protein Folding, 68
Ion Chromatography, 110-112, 114-116, 119-121, 131, 197
Ion Cyclotron Resonance, 78, 158-161, 163, 167
Ionic Forces, 73
Isoelectric Point, 16, 28, 34-35, 79, 97, 100-101, 108, 118, 180
Isomerism, 31

L
Label-free Quantification, 84-85
Levinthal's Paradox, 69
Ligand Binding, 14
Liquid Chromatography, 72, 79-80, 131-132, 153-154, 168, 190, 192-193, 196-202

M
Magnetic Quadrupole, 151-152
Mass Analyzer, 79, 146, 153, 156-160, 162-164, 168, 170-171
Mass Spectrometre, 153

Index

Mass Spectrometry, 3, 5, 7, 18, 25, 49, 74-85, 103, 109, 132-133, 138, 141, 143-144, 148, 153, 158-164, 167-170, 181, 187-189, 200, 223
Membrane Exchange Chromatography, 117

N
Neurodegenerative Disease, 68
Nullomers, 41

O
Optical Tweezers, 68

P
Partition Chromatography, 29, 72-73, 192-193
Peptide Bond Formation, 42
Peptide Fractionation, 79
Physicochemical, 41, 45
Polar Forces, 73
Preparing Acrylamide Gels, 93
Primary Structure, 11, 37, 49-50, 56, 81, 97, 172, 174, 178-179
Protein Breaking, 29
Protein Complementation Assays, 6
Protein Disorder, 20
Protein Extraction, 23, 27, 105
Protein Fold, 52, 56
Protein Identification, 76, 80, 83-85, 107, 154, 169-170
Protein Nuclear Magnetic Resonance Spectroscopy, 67
Protein Preparation Protocol, 22
Protein Purification, 16, 110, 117, 125, 205
Protein Quantitation, 81
Protein Sequence Analysis, 53
Protein Structure, 1, 7, 13, 19-20, 22, 32, 41, 45, 47-49, 51, 53, 55-56, 58, 62, 64, 82, 92, 172
Proteogenomics, 83
Proteolysis, 23, 53, 68, 131, 180
Proteomic Analysis, 2-4, 26-27, 76, 84, 223
Proteomics Study, 1

Q
Quadrupole, 78, 146, 149-153, 157, 159-160, 162-163, 187, 189
Quantitative Proteomics, 75, 83, 103, 167
Quaternary Structure, 12, 51, 58, 66, 129, 198

R
Rehydration, 107-108

S
Sample Fractionation, 24
Sample Preparation, 22-23, 27, 78, 85, 92, 105, 107, 120
Secondary Structure, 11, 50, 52-53, 57-58, 61, 63, 179
Separating Proteins, 117
Sequence Motif, 51
Side Chains, 8, 10, 29-30, 32-35, 41-42, 45, 50, 60, 66, 118, 165, 172, 176-177
Strong Ion Exchangers, 115
Structural Proteins, 13, 15, 131
Structure Determination, 5, 12, 32, 53, 82
Superdomain, 52
Supersecondary Structure, 52

T
Targeted Proteomics, 84
Tertiary Structure, 10-13, 50-51, 53, 58, 64, 129, 179-180, 198
Trace Amines, 39
Triple Quadrupole, 157, 160, 162

V
Vibrational Circular Dichroism, 67

X
X-ray Crystallography, 6-7, 12, 48, 53, 65, 86

Z
Zwitterions, 33-35

Printed in the USA
CPSIA information can be obtained
at www.ICGtesting.com
LVHW082022140823
755210LV00007B/550